A COMPENDIUM OF

ARTISANS, TRADES & PROJECTS

FOR THE MAKERS

A COMPENDIUM OF

ARTISANS, TRADES & PROJECTS

VANESSA MURRAY

hardie grant books

CONTENTS

INTRODUCTION

The most beautiful gift I've ever received is a bedside table. It has four Louis XV– inspired legs and a single drawer that sits neatly atop a single shelf, where my magazines fit perfectly. It's crafted from Queensland Maple, Silky Oak and Tasmanian Blackwood timbers, and is held together by dovetail and mortice and tenon joints. The drawer has a pull that was handcrafted from moonstone and silver by a dear old friend of mine who is a goldsmith in New Zealand.

It's almost as though the table was made for me – and in fact, it was, by the hands and heart of the man I love. It took him more than 50 hours, with a lot of sanding, and a little bit of swearing. I imagine that our son will inherit it when we're gone, and eventually it will sit beside the bed that he shares with the one he loves.

Not long after receiving the table, I started working on this book: *Made to Last: A compendium of artisans, trades & projects*. I was inspired by my growing interest in the provenance of things, and that of people around me, by my talented partner's desire to make furniture as beautifully and as well as he possibly can, by friends' desire to buy cookware or garden tools or lights that will last for as many years as they want to use them, and by the emergence of websites that only sell items that come with lifetime guarantees.

This interest has grown out of a deep dissatisfaction with the disposable nature of the world we live in. The XYZ is broken? Time to get a new one. The ease with which we can replace things discourages us from caring for the things we buy, or from investing a little more and buying something that will last longer.

Not so many generations ago, the tools of everyday living were all made by hand, and were treated with practical reverence by their makers and their owners. Take your kitchen knife. It will serve you best when it's clean and sharp and dry, but it won't stay like that on its own. In order for the knife to serve you well, you must serve the knife; it is you who must keep it clean and sharp and dry.

It's a reciprocal kind of service – I look after you, and you look after me. It's simple, really, and I feel optimistic that we are turning away from our throwaway culture and moving into a time where we treasure our everyday items once again.

In *Made to Last*, I set out to find and learn about artisans from all around the world who are making useful, heirloom quality objects that will stand the test of time. Some are practising a trade that has been their family's lifeblood for centuries; others have determinedly read and watched and trialled and errored, and taught themselves.

What is this phrase, 'heirloom quality'? I was inspired by inventor Saul Griffith's notion of 'heirloom design', which he applies to objects that last through the original purchaser's lifetime and into the next generation, because of the inherent robustness of both their form and their function. Products with heirloom design value are physically well made and highly functional, but aesthetically beautiful, and timeless too.

Many of the items in *Made to Last*, if cared for properly, will be handed down through generations. The Blackwood rod back settee made by Glen Rundell, for example, or Jeremy Zietz's Campaign chest. Items like these have been made to the highest standards with love and care and little regard for modern manufacturing's interest in profit margins, or productivity at any cost.

Others, like furniture maker Shigeki Yamamoto's Eternal calendar and glassmaker Nate Cotterman's Cube Glass, are beautifully and thoughtfully made, but also engage playfully with the book's title and themes. Still others, like Re'em Eyal's Derby lamp or Dominic Odbert's portable sound system, the Wood Grain BoomCase, take junkyard finds and breathe new life into them with clever upcycling.

Not everything in *Made to Last* will be handed down in physical form to a child or grandchild. Take the fermented foodstuffs. Their shelf life is long, but they won't last a hundred years. In these cases, it's the knowledge that has heirloom value, rather than the goods themselves. Will you teach your son how to keep a sourdough starter alive, and how to use it to make bread for himself and his family? Will your show your granddaughter how to make her own fire cider, or her own bathtub gin? I hope so.

Each product story in *Made to Last* is really the story of its maker. When and where and how did they learn their trade? What inspired them to take this path? What tools do they use to make the things they do, and how do they feel while they're making them?

The makers you'll meet here share a love of their craft, and a curiosity about how things work, and how they might be made to work. They are resolutely working against today's throwaway culture, one beautiful object at a time.

I've included product care advice and, where possible, a DIY-style project to try at home, inspired by the maker's creative process. This might be an entire project, like industrial sewer Cathy Parry's 'How to make a bike pannier', or a transferrable skill from a process, like wheelwright Mike Rowland's step-by-step guide to 'How to make a mortice and tenon joint'. In other cases, as with horologist Stephen McGonigle's Ceol minute repeater wristwatch, we've simply shown how a particular part of the making process happens.

One thing you might notice missing from the stories is an indication of what the goods I've chosen to showcase cost. There are a few reasons for this. One is that prices change over time. Another is that this is not a shopping catalogue. Look, be inspired, fall in love ... Visit a maker's website and find out what this or that will cost you. Some things will be instantly affordable; others you might spend months or years saving up for.

Finally, just as the trades and skills being shared here are timeless, the goods these makers produce are priceless.

Many of the items in *Made to Last* are being made in the same ways they were 100 or 500 or 2000 years ago. And you can't put a price on that.

JUKI

JUKI
JUKI

SHOP APRON

Craft sewing

If you plan to get dirty but want to look good while you're doing it, you'll need an apron. But not just any old apron – a sturdy, fit-for-purpose, extremely well thought out and made apron. This shop (that's 'workshop' to British English speakers) apron is just the thing. It was made with equal parts love and determination from waxed, 10.1-oz army duck canvas by a lifelong craft sewer who lives in the countryside in Oregon, USA, who practises her craft in a converted trailer that's permanently parked in her backyard. The shop apron is suited to art teachers, gardeners, woodworkers, blacksmiths, mechanics, instrument repairers and pretty much anyone who wants to get stuff done. It has 14 pockets, for tools, keys, phones, snacks, pens, papers and calling cards, is water-repellent, and will only improve with age and wear.

MEASUREMENTS	Made to order
MATERIALS	283-g (10-oz) army duck canvas with martexin original wax, bonded nylon thread, nylon flat weave webbing, nylon grosgrain ribbon, plastic hardware (side release buckle, triglides, ladderloc buckles, elastomer tip ends, industrial snap)
KEY TOOLS	Chalk pen, double-pointed knitting needle, extra-long quilting pins, knife-edge tailor's shears
KEY MACHINES	Edge hot knife, industrial bartack machine, industrial needle feed sewing machine, professional snap press
TIME TO MAKE	2 hours
LIFESPAN	Heirloom quality

RANDI JO SMITH — Craft sewer [Oregon, USA]

The first apron Randi ever made was a birthday gift for her husband. He was working as a mechanic at the bicycle shop they owned at the time, and she couldn't stand seeing him wearing branded polyester aprons that weren't handling the grease he was throwing at them. So she set about making him an apron that was far more durable, worthy and handsome.

That was back in 2006, when Randi and her husband were courting. Today, the pair live on a 15-acre (6 ha) farm in the Oregon countryside with their two children. They've sold the bike shop, and Randi Jo Fabrications has become the family's sole source of income. In addition to aprons, Randi makes cycling caps, handlebar and tool bags, waterproof saddle covers and caps and bags for kids.

Randi's workshop is just 20 steps from her back porch, in a 3 × 6.7 m (10 × 22 ft) travel trailer that her father converted for her.

'It has a shed roof, a single door and plenty of windows. Essentially a tiny home on wheels with an open floor plan, it's perfectly suited for my sewing business,' says Randi, who can see her garden, complete with browsing goats, chickens, peafowl, ducks and wild turkey, from her trailer.

'I learned to sew when I was eight or nine. My mother, grandmother and great-grandmother are all handy with the sewing machine and each of them have taught me different aspects of the craft.'

Randi makes shop aprons in small batches of two or three at a time. She begins by taking the pattern, which she created herself, marking up the army duck canvas with a chalk pen and cutting it with knife-edge tailor's shears. Then she cuts the nylon webbing for the shoulder and waist straps with a hot knife.

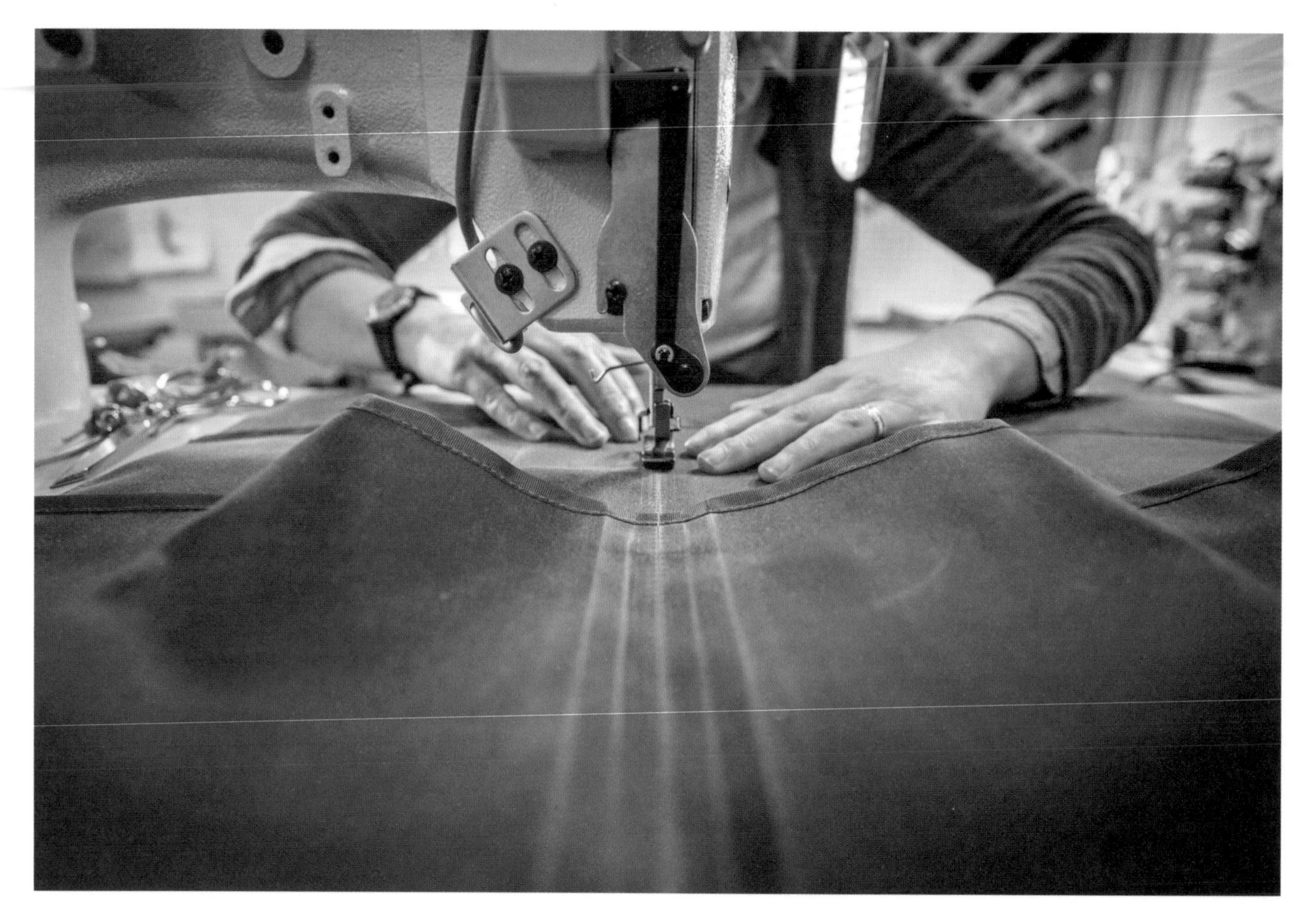

'I sew the hem of the tiny chest pocket and turn it right side out with the double pointed knitting needle. I sew the tiny pocket to the main chest pocket, then I sew that whole assembly onto the apron body using my industrial needle feed sewing machine.'

Randi sews nylon grosgrain ribbon along the tops of the lower pockets, again on the industrial needle feed sewing machine, and then marks the fold lines and sewing lines with the chalk pen. She sews the lower pockets onto the apron body using the needle feed sewing machine. The bottom row of pockets is pleated and sewn after the first flat pockets are attached. Once all of the pockets are attached, she makes the waist strap assembly and attaches it to the sides of apron using the 'box X method' on the needle feed sewing machine.

She sews nylon grosgrain to the outside edges of the apron to enclose the raw edges of the waxed canvas. The top of the apron is folded down to form a hem and the shoulder webbing straps are enclosed in the hem. The shoulder webbing straps are then folded up and secured – again, with the box X method on the needle feed machine.

Now Randi bartacks (applies reinforcing stitches) each apron at its stress points (at every pocket seam) using an industrial bartack machine. She slides a triglide buckle onto each shoulder strap (this enables them to be adjusted) and attaches the ladderloc buckles at the waist point. The straps are finished with the elastomer tip ends. Lastly, she adds a snap to the apron body under the chest pocket, using the professional snap press.

'My favourite machine is the industrial needle feed machine. I got it in 2016, and it makes my job so much easier and faster; it has an automatic backtack and thread cutter and it sews such consistent and lovely seams. My favourite hand tool might just be that double pointed knitting needle. I use it to turn pockets and it helps make that tiny chest pocket on the apron perfectly square. I get a little panicky when I can't find it!'

CARING FOR YOUR SHOP APRON

Do not wash the apron. Simply spot clean it as necessary, with a damp cloth. The aprons get better with lots of build-up ... If the build-up starts to bother you, however, reapply warmed wax with a little elbow grease.

"

My favourite hand tool might just be that double pointed knitting needle ... I get a little panicky when I can't find it!

MEGAMAT

HOW TO

MAKE AN APRON AND TOOL ROLL

MATERIALS

- Canvas – 283 g (10 oz) (denim or another heavy weight fabric will also work)
- 2.5 cm (1 in) nylon grosgrain ribbon or twill tape
- Thread
- 2.5 cm (1 in) webbing
- 2.5 cm (1 in) side release buckle
- Grommet
- Leather or nylon cord

TOOLS

- Scissors
- Sewing machine
- Ruler
- Chalk or pencil
- Pins
- Grommet press

METHOD

1. Cut a canvas rectangle 53.3 × 35.6 cm (21 × 14 in) for the main apron.
2. Cut another rectangle 53.3 × 17.8 cm (21 × 7 in) for the pocket.
3. Mark the pocket at 12.7 cm, 20.3 cm, 33 cm and 40.6 cm (5 in, 8 in, 13 in and 16 in) vertically to make five pocket compartments. You can experiment with the pocket sizes depending on the end use of the apron.
4. Sew the grosgrain ribbon or twill tape to the top of the pocket piece by folding the ribbon in half lengthwise and encasing the pocket piece inside the ribbon.
5. Place the pocket piece on top of the main apron, lining the bottom of the pocket up with the bottom of the apron. Stitch around the edges of the pocket.
6. Stitch the pocket compartments you marked previously and be sure to backtack or bartack the end of the seam to secure it.
7. Sew the grosgrain ribbon around the outer end of the apron in the same manner you did on the pocket. Randi suggests cutting each corner of the apron into a curve so you can easily wrap the ribbon around the edge to make this easier.
8. Cut a piece of webbing long enough to go around your waist, plus 15.2 cm or more (6 in +) to enable adjustment with your buckle.
9. Sew the buckle hardware on each side of the webbing.
10. Mark a placement line for your webbing on the apron 5 cm (2 in) above the pocket.
11. Place the webbing above the placement line and stitch it to the apron, starting at the edge of the apron just inside the ribbon. Sew the top and bottom of the webbing to the apron.
12. Mark where the grommet will be placed, approximately 10.2 cm (4 in) from the bottom of the apron and directly next to the ribbon.
13. Attach the grommet.
14. Cut a piece of leather or cording approximately 61 cm (24 in). Fold it in half and pull it partway through the grommet. Pull the end of the cord through the loop in the cord that is through the grommet.
15. To fold the apron into a tool roll, fold the top section (above the webbing) over the pocket section. Now you can roll or fold the apron starting at the end without the grommet. Use the cord to tie your bundle and you're ready to roll!

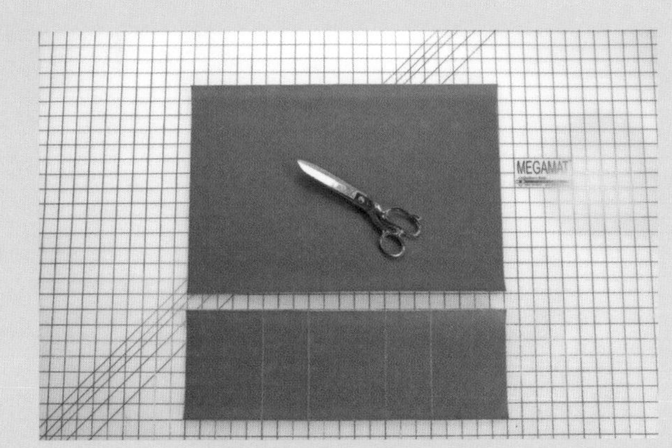

Steps 1–3

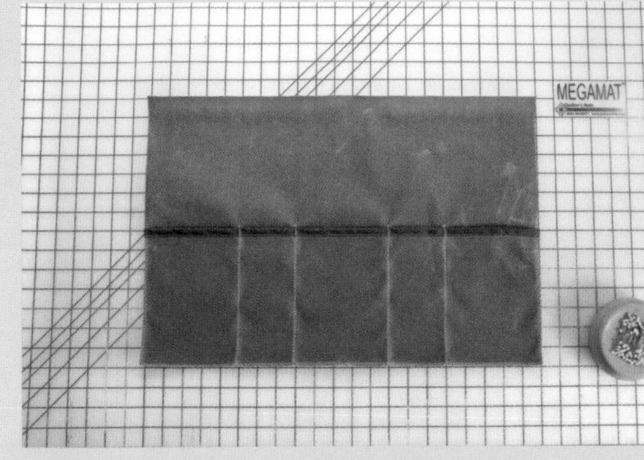

Step 5

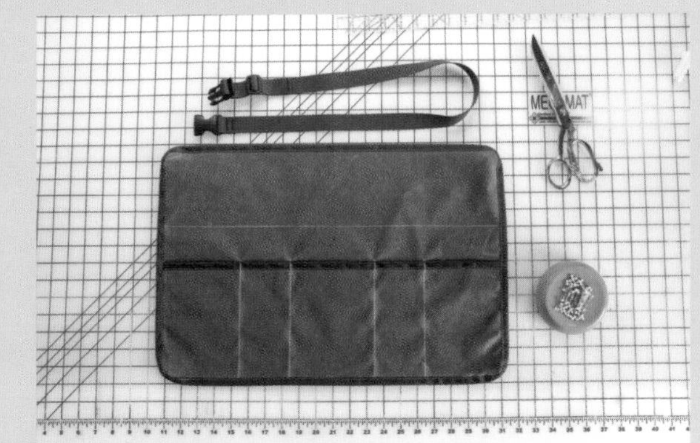

Steps 7–8

Step 15

W
M
WOLF &
MAIDEN

WOLFPACK BACKPACK

Leathersmithing

Three cheers for this roguish number, which enables you to carry everything you need for the day, yet still travel empty handed. It's called the Wolfpack, and it's crafted from a single piece of leather, sourced from a fourth generation tannery in Zululand, South Africa, that adheres to vegetable tanning techniques of old. This backpack is meticulously stitched and stained by hand, then embellished with brass, in a workshop in Cape Town. Its dividing partition cleverly doubles as a structural feature, and the backpack is drawn together using the time-honoured technique of saddle stitching. Perfected by master guildsmen in medieval times and passed down through generations, saddle stitching ensures optimum reliability and strength. Like everything that comes out of Wade Ross Skinner's workshop, Wolf & Maiden, the Wolfpack is designed to improve with age – and use.

MEASUREMENTS	Weight 25cm × height 40cm × depth 12 cm (10 × 16 × 5 in)
MATERIALS	Chicago screws, leather treatment, solid brass holster studs, stain, vegetable tanned leather
KEY TOOLS	Bone, burnisher, edge cutter, hammers, knives, needles, punches, rulers, stitching pony
KEY MACHINES	Spray gun
TIME TO MAKE	8 hours
LIFESPAN	Heirloom quality

WADE ROSS SKINNER — Leathersmith [Cape Town, South Africa]

It was a chance encounter in India in 2011 that inspired Wade Ross Skinner to learn leathercraft. Well, two chance encounters ... Twice, on opposite ends of that sprawling subcontinent, Wade crossed paths with a mysterious French leathersmith. Each time, he assisted the man with food and shelter. In return, the Frenchman fashioned him an intricate leather tobacco pouch with an almost mystical air.

On his return to Cape Town in 2011, Wade resolved to learn how to work leather into beautiful shapes for practical purposes, and Wolf & Maiden Creative Studio was born.

Wade was a bedroom crafter initially, holed up in his share house from dawn 'til dusk, honing his craft. The first thing he made was a pencil case, and it was far from perfect. Within three weeks, he'd driven his housemates mad with the sound of hammering, but he'd made something he could sell.

These days, Wolf & Maiden employs more than 30 staff, and operates out of an expansive former synagogue in Cape Town. Visitors enjoy windowed views into the workshop, and can continue through to a bar and cafe if they choose. Wade wants people to linger, to immerse themselves in what he calls 'heritage crafting'. He trains all his craftspeople, and prototypes everything himself.

'I don't get hands-on as much as I used to, but I try to get creative often. When I'm in the workshop time doesn't exist, my own thoughts don't exist. I run on autopilot. It's time to quiet the ego and mind and plug into creative channels. The trick these days is creating the time and space to do that,' says Wade.

The Wolfpack is one of the more complex pieces in Wolf & Maiden's collection of around 30 beautiful and practical items. Wade uses one himself – a black one – every day.

He developed the idea after cutting a piece of leather to around the size of the backpack he had in mind, then bending it this way and that, observing how the leather behaved. He made notes about measurements and other materials, then got to work on the template, drawing it up on cardboard.

To make a Wolfpack, the template is laid onto the flesh side of the leather and marked up with a silver leather pen.

'The templating must be spot on,' Wade says. 'It's tempting to rush it, but when you lay everything out by hand and mark every hole by hand ... you just need to miss one detail for the final product to be out of kilter.'

When he's satisfied, Wade places the leather on a corkboard and cuts the shape out using a steel ruler and leather knife. Next comes hole punching for the stitches, shoulder strap and remaining hardware – a holster stud and screws at the base, to stop the sides from sagging.

He runs an edge cutter along the edges to smooth out the corners, then the leather and the edges are stained. Wade puts on protective gloves, a mask and goggles, then sets to work with a cloth or a spray gun. Multiple applications are performed back to back then left to dry. Because the stain is done by hand, no two bags are ever quite the same.

Once the leather has dried, it's time for assembly and stitching. Wade uses a tool called a bone to mark and bend the seams, then folds the Wolfpack's single piece of leather in upon itself, and places it in the stitching pony. This leaves both his hands free for the labour-intensive saddle stitching. Stitching the entire Wolfpack takes about five hours. Burnishing comes next – any flesh edges that meet are burnt so that they don't fray. Then comes hardware, and then a top-secret leather protector treatment, finish and buff.

'I tell people that the day their bag leaves our store, it looks its worst. Vegetable tanned leather is incredible stuff – it's continually absorbing oils and moisture; it just gets better with age. I love that people who buy my work are going to go on and customise the piece themselves, simply through everyday use.'

CARING FOR YOUR WOLFPACK

Vegetable tanned leather requires very little care. It will improve over time, softening and becoming more supple with use. It will absorb oils from the user's skin and handling, so the parts of the bag that have more skin contact, such as the straps and flap, will soften sooner. It is hand-stitched, which means that if the thread does eventually come undone, it can simply be restitched by a keen home saddle stitcher or a professional leather worker. The one thing to avoid is saturation. The Wolfpack can withstand a rain shower or two, but if you – and it – get stuck in a thunderstorm, the leather will become brittle and dry. If this does happen, stuff the Wolfpack with newspaper and wait for it to dry naturally, then give it a decent moisturising leather treatment, stat.

“

When I’m in the workshop time doesn’t exist, my own thoughts don’t exist. I run on autopilot. It’s time to quiet the ego and mind and plug into creative channels.

HOW TO

MAKE A KEY RING

You can customise this piece by sourcing leather from your area, by sourcing unstained leather and staining it yourself using stains from plants native to your area, or by punching/marking additional patterns into the leather.

MATERIALS

- Cardboard
- Piece of leather ~ 10 × 10 cm (4 × 4 in)
- Key ring
- Waxed linen thread or M36 thread

TOOLS

- Ruler
- Pen
- Craft knife or scissors
- Cork board (or a chopping board you don't mind marking)
- Leather blade or sharp craft knife
- Multi punch
- Mallet
- Cloth
- Stitching pony or vice (optional)
- Leather needles × 2

METHOD

1. Use the template in the image as inspiration for your own design, then use the ruler and pen to sketch your template on the cardboard and cut it out using the craft knife or a pair of scissors.
2. Lay your cardboard template on the flesh side of the leather, then trace the outline onto the leather using your pen.
3. Pressing firmly, use the ruler and leather blade or craft knife to cut out the shape.
4. Use your pen to mark the holes where the stitches will go, then take the multi punch and mallet and punch out the holes.
5. Place the key ring at the mid-point and fold the leather in half so that the holes match up, then place the cloth around the leather, leaving the area to be stitched exposed.
6. Put the leather in the stitching pony or vice and tighten it up, ensuring that the pony only comes into contact with the cloth, not the leather. The cloth is to protect the leather from scratches while it's in the pony. (You can complete this project without a stitching pony or vice, but having something to hold the leather still while you stitch it really helps.)
7. Cut two lengths of thread about 30 cm (12 in) long, then thread both of the needles. Now take one of the needles and push it through the first hole in the row you're stitching. Pull it about halfway through, then take the second needle and push it through the same hole, from the other side.
8. Proceed to stitch along the row, leading with the left-hand needle then the right-hand needle. Both sides of the leather will be stitched, and the needles will 'swap sides' with every stitch. Take care to pull the thread tight at the end of each stitch so the stitches sit in an even line.
9. Tie off at the end by doubling the last stitch, then tying a tight knot and cutting the remaining thread with your scissors, close to the knot.
10. Stitch the other seam in the same fashion.

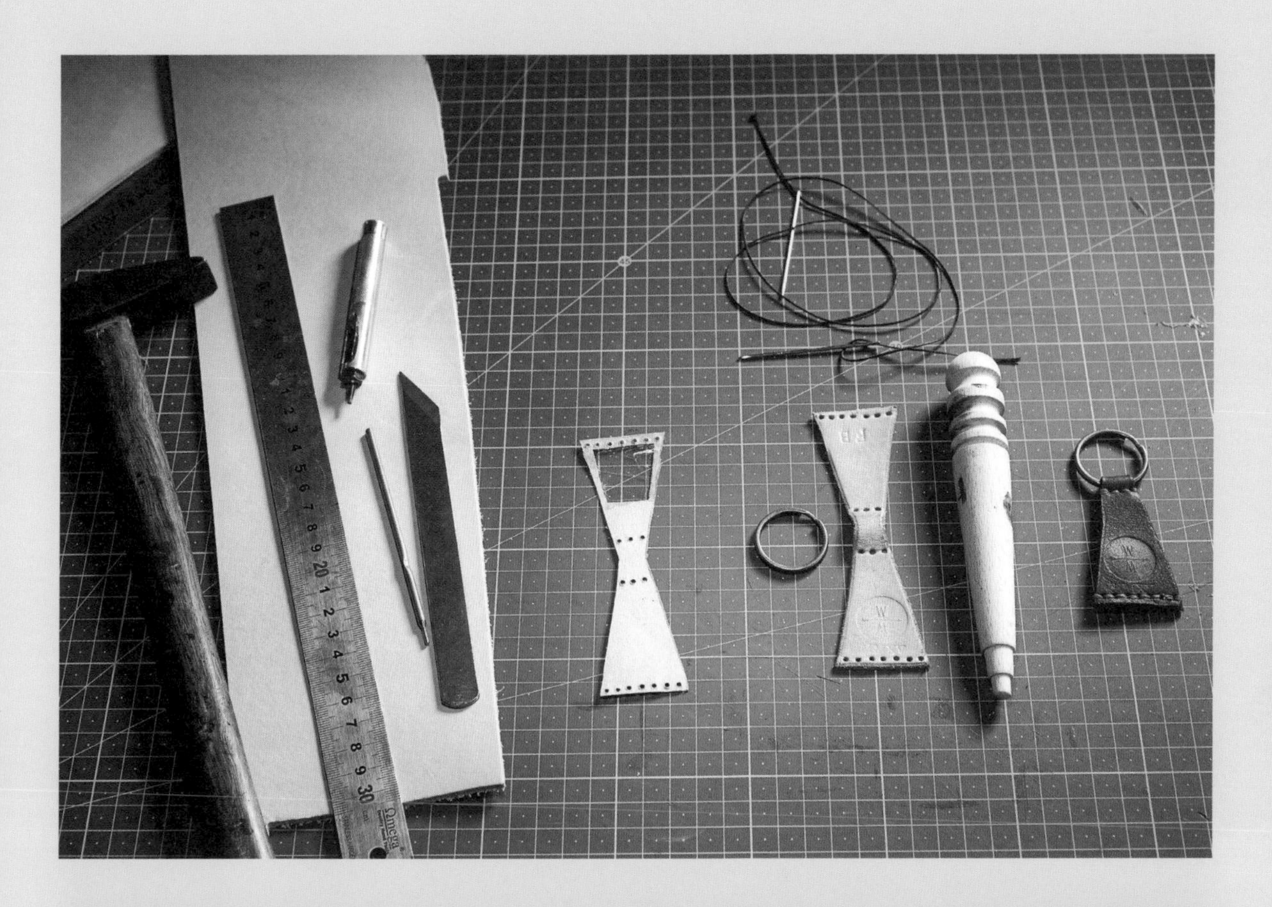

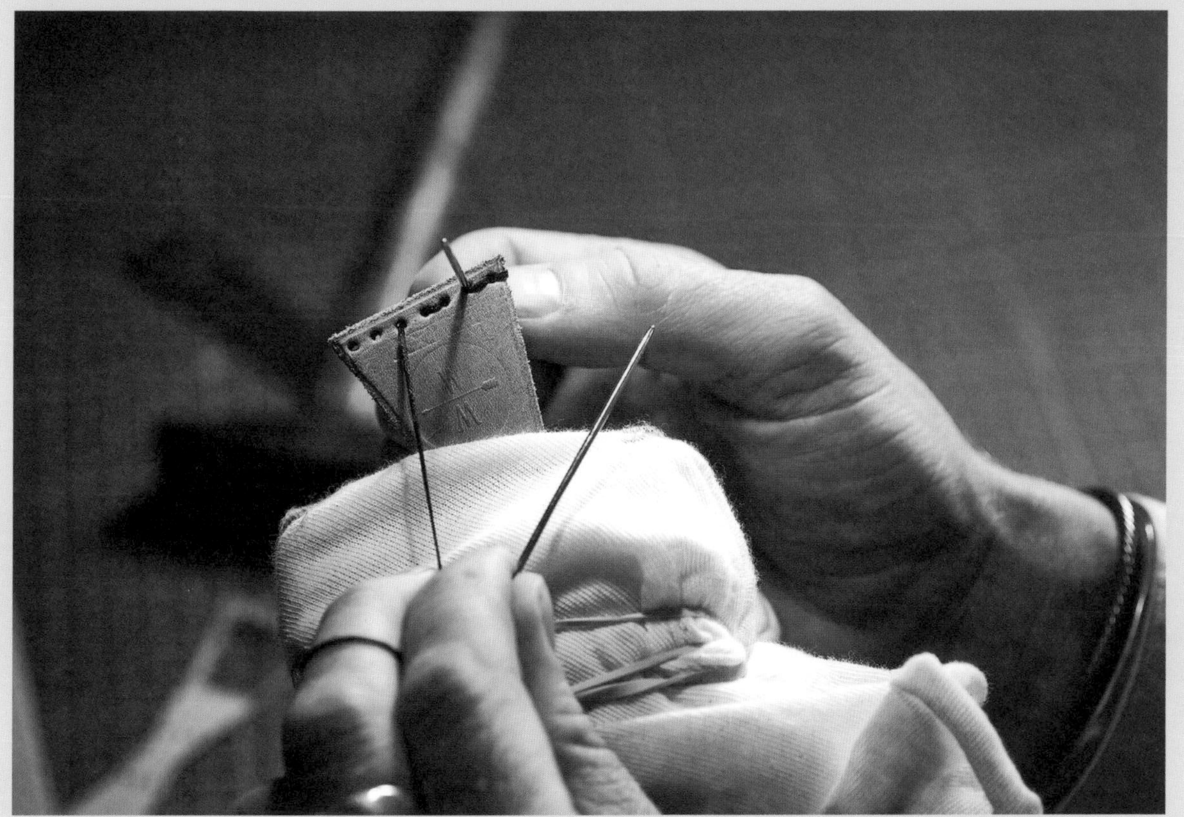

Steps 7–8

百货
食品
饮料

BAMBOO BICYCLE

Bicycle making

Bamboo isn't just for pandas. As one of the world's fastest growing renewable resources, with the lightness and strength to rival materials like concrete and aluminium, bamboo is becoming popular with all sorts of designers and makers, too. And, as the crew at DIY maker space Bamboo Bicycles Beijing (BBB) has discovered, it's perfect for bicycles. Light, hollow and stiff enough to withstand the rigours of cycling, bamboo has fancy unidirectional fibres that mean it can handle loads of tension and compression, while its suppleness doubles as an excellent shock reducer. Hemp rope has been wrapped around the lugs to complement the bamboo's clean, organic vibe. This particular model is a single-speed road bike with caliper brakes that BBB founder and chief tinkerer David Wang made for a friend over the course of a weekend.

MEASUREMENTS	Frame: 54 cm (21 in); Weight: ~ 3 kg (6 lb 10 oz)
MATERIALS	Bamboo, carbon fibre, electrical tape, carbon fibre tow, sandpaper, tacking glue, epoxy resin
KEY TOOLS	Assorted bike tools, hacksaw
KEY MACHINES	Dremel handheld rotary tools, drills, hole saw, spindle sander
TIME TO MAKE	2 days
LIFESPAN	5 years +

DAVID WANG — Bicycle maker [Beijing, China]

Thirty years ago, Beijing was the bicycle capital of the world. Today, as a result of decades of industrialisation and motorisation, the city is filled with smog, and bicycles are in short supply. This, along with a love for all things bike-related, motivated David Wang to make his own bamboo bicycle, and to launch a community workshop where locals can do the same.

There's room in BBB's workshop for five people to get hands-on at any one time. The workshop is located in a former 24-hour mahjong den in Beijing's Dongcheng District. It was launched with the help of a Kickstarter campaign in 2014, and it's the first do-it-yourself bamboo bicycle workshop in China. Since its inception, more than 400 bikes, from road bikes and mountain bikes to tandem and sidecar bikes, have been made within its 80 sq m (861 sq ft).

The first thing David does when he's starting on a new bike project is to design the frame to fit the rider, using custom bicycle design software RattleCAD. Then he prepares the jig (an adjustable aluminium frame that he made himself) that will hold the pieces of bamboo in place as they are put together.

After he's measured and marked up the bamboo, he cuts it using a hacksaw, then shapes the ends for joining. It's one of his favourite parts of the entire production process. 'There's something relaxing and rewarding about making two pieces of bamboo fit snugly together. I think this is where craftsmanship really comes into play. If I'm too impatient, it's almost impossible to get things to fit, but if I take my time everything seems to flow.

'I carefully shave down the ends of the bamboo so that it fits at exactly the correct angle and diameter of the other tubes it is joining. I used to do the whole thing using a handheld rotary Dremel, and I got pretty good at it. But these days in the BBB workshop we've got better tools for the job: a hole saw and – my favourite – a spindle sander.'

After prepping the various pieces of bamboo for tacking together by roughing up the ends with sandpaper, he glues them together in the jig, in the following order: seat tube, down tube, top tube, right chain stay, left chain stay, right seat stay, left seat stay. After it comes out of the jig, he moves on to creating the bottom, head and seat lugs – the three joins that hold the different parts of the frame together. This involves carbon fibre, safety gloves and goggles, and a lot of extremely adhesive epoxy glue. It's the trickiest part of the whole build.

'Wrapping carbon fibre by hand is a headache and a half! You need to get all dressed up with protective gear and then mix sticky epoxy. Then you need to mix that sticky mess with carbon fibre and begin wrapping in a bunch of different directions. It's incredibly tedious, and it's nearly impossible to do a flawless wrap with so much epoxy everywhere.'

The wrapping goes a lot better working with a friend, says David. Once it's done, the frame needs to be left to cure. After a few hours has passed, the tape and plastic wrap can be removed, and the other bike parts can be installed.

David has always had a love of bikes. He grew up in the United States, and learned to ride when he was six.

'I wasn't particularly creative as a kid, but I really liked being independent. I've been working in China for about ten years. I started as an anthropological researcher, but the vestiges of Beijing's bike culture piqued my interest a few years ago. I began salvaging old Beijing bikes from the 1980s and before I knew it I was making a bamboo bike. I love being able to understand something and then reproduce it on my own. Bicycles really lend themselves to that.'

"

If I'm too impatient, it's almost impossible to get things to fit, but if I take my time everything seems to flow.

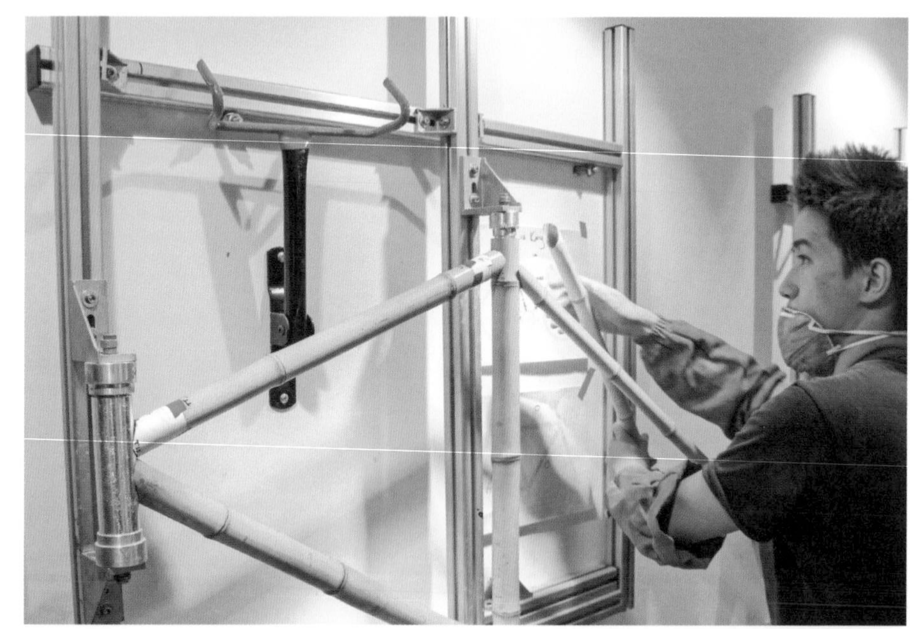

CARING FOR YOUR BAMBOO BICYCLE

The key to preserving your bamboo frame is to waterproof it. This is done easily by applying a generous amount of hardwood floor wax to the frame once or twice a year, as winter (or your wet season) approaches. Your bamboo frame should outlive your bike components by a long shot.

HOW TO

MAKE A BAMBOO BICYCLE FRAME

This DIY assumes you know the measurements for your frame and that you have all the other parts you'll need for your bike. Oh, and that you know how to install them.

MATERIALS

- 3 thick bamboo poles, 30 mm (1.2 in) diameter, for the top tube, down tube and seat tube
- 4 short poles, for the left and right seat stays and the left and right chain stays, 20 mm (0.8 in) each
- Plastic wrap
- 10 packets of tacking glue
- 12 fibreglass strips
- 30 spools of carbon fibre tow
- 1.5 kg (3 lb 5 oz) epoxy resin
- Electrical tape

TOOLS

- Pencil
- Vice
- Hacksaw and blades
- Dremel
- Frame jig (aluminium is easiest)
- 80-grit sandpaper
- Tray for soaking fibreglass strips
- Glue brush
- Craft knife

SAFETY GEAR

- Goggles
- Latex gloves

METHOD

1. Prepare your jig to the dimensions of your desired bicycle frame by adjusting its vertical and horizontal bars.
2. Measure the bamboo according to the dimensions of your frame, and mark it up with pencil.
3. Cut the bamboo by placing each piece in the vice and use the hacksaw to carefully cut along the edges you have marked up.
4. Gradually mitre the bamboo until the joints come together snugly inside the jig. Each end needs to be bevelled at a 45-degree angle to become smooth and tapered; do this with your Dremel set to 3-speed or higher.
5. It's time to prepare the bamboo for tacking and gluing. Lightly rub the ends of each piece of bamboo with sandpaper to roughen it up.
6. Tape a layer of plastic wrap over each piece of bamboo, leaving a palm's length at either end.
7. Put on your safety goggles and glasses, have your tray, glue brush and craft knife handy, then prepare the tacking glue according to the instructions, then glue the pieces together. Leave the glue to dry, again as per the instructions for the glue you're using.
8. Now that the bamboo is in the shape of your frame, you can remove it from the jig and set about reinforcing each joint using the fibreglass strips, carbon fibre tow and epoxy resin. It's vital that this part of the process is done safely, and under the supervision of an experienced bamboo frame maker. At BBB, each joint is subject to a specific wrapping pattern that has been designed to cope with the different loads riding will demand of that part of the frame. Each wrapping pattern is repeated seven times. Seek out an expert, or some expert information, to guide you here.
9. Leave your frame to cure for as long as the epoxy brand you're using specifies, and be sure to remove the plastic wrap before installing your bike parts.

Step 2

CHEF'S KNIFE

Bladesmithing

This knife was made entirely by hand by a largely self-taught bladesmith and blacksmith. It is a chef's knife of the highest quality, with a handle of red mallee burl timber and a blade of Damascus steel, which is renowned for being tough, resistant to shattering and able to be honed to a sharp, resilient edge. Damascus steel has its origins in the somewhat hazy history of the Ottoman Empire, and was the steel used for the blades of its warriors. At the height of their powers in the 16th and 17th centuries, the Ottomans controlled much of south-east Europe, and parts of central Europe, western Asia and Africa. No records remain to tell us exactly how they made their steel, but, like today's, we know it was forged by layering and joining different types of steel together, and typically had a beautiful pattern reminiscent of water eddying in a pool.

MEASUREMENTS	Total length: 36 cm (14 in); blade length: 23 cm (9 in); handle length: 13 cm (5 in); blade width: 4.5 cm (2 in); weight: 200 g (7 oz)
MATERIALS	1075 high carbon steel, 15N20 high carbon steel, red Mallee burl timber, stainless steel
KEY TOOLS	Anvil, hammers, leather apron, tongs
KEY MACHINES	Belt grinder, forge, fly press, hydraulic press, oxyacetylene torch
TIME TO MAKE	Damascus steel: 1 week; knife: 25 hours
LIFESPAN	Heirloom quality

IAIN HAMILTON — Bladesmith and blacksmith [Dignams Creek, Australia]

Damascus is the most intricate type of steel that Iain Hamilton creates at Mother Mountain Forge, which he founded in the foothills of New South Wales' south coast in 2012. He makes blades of all types to order, to fit the hand of their intended user, which means no two of his knives are ever the same. And they're beautiful.

'But no matter how beautiful or intricate it is intended to be, a knife is still a tool. All knives must be practical and useable, so that's my first priority,' says Iain, who also handcrafts all his handles and makes his own sheaths, scabbards and display boxes.

Forging Damascus steel is a highly intensive process. It can take up to a week, depending on the complexity of the pattern Iain has in mind.

Iain stacks different steels, tack welds them to a rod, then covers them in a flux (a wet weld application of anhydrous borax), which stops oxygen getting in between the layers. He heats it in the forge to 1200°C (2192°F), then gives it a strong hammering on the anvil to weld the layers together, forming what's called a 'billet'.

'My anvil is well over 100 years old, so carries the energy and history of the craft within it. It's also the one I learnt on. Along with my cross pein hammer, it's one of my favourite tools.'

Iain inspects the billet. If it's not all welded he adds more flux, reheats it in the forge and hammers it again.

He repeats this process until it's done, then lengthens it using the hydraulic or fly press, reheating it in between drawings so it remains malleable. He then cuts the billet into equal pieces, restacks it and begins the process all over again, until he has the number of layers he's after.

'After that I pattern the billet. For this knife I drilled into the billet, then reheated and hammered it so that the circles flattened out and the layers were visible, like raindrops.'

Once Iain has crafted the Damascus steel, he gets to work turning it into a knife blade. This time he heats the metal to 1000°C (1832°F) – he knows it's right when it turns a bright orange colour. Then he uses the tongs to take it to the anvil, and hammers it into shape.

Next, Iain normalises the blade, which means heating it in the forge to a non-magnetic heat, then cooling it by swinging it through the air. Then he anneals it by heating the blade until it's a blood red colour (around 600°C/1112°F) and cooling it slowly at room temperature. This relaxes the steel, making it soft and ready for grinding. After grinding comes heat treatment, then a series of polishing grinds on the belt grinder using progressively finer grits. Iain does a final polish of 1200 grit by hand.

Then it's time to make the handle. Iain selects and cuts the timber, marks the centre at one end and drills a pilot hole. He uses an oxyacetylene torch to heat the tang – the end of the blade – and burns it into the block. Then he shapes it to fit a ferrule – a stainless steel collar – on the belt grinder.

'With a Damascus blade, this is when it is acid dipped to reveal the pattern. I submerge it into ferric chloride acid for 15–30 minutes then wash and rinse it, and give it a light hand sand with 1200 grit wet and dry sandpaper.'

Iain drills the handle block and tang to fit a pin, then glues and pins the block and ferrule in place. He sands them, treats them with a mixture of beeswax and linseed oil, and sharpens the blade. The knife is now finished and ready for use.

'I love creating things that will last for generations. Forging steel into a warm, natural material rather than a cold, man made substance links me to the past; I am working it the way it has been worked for thousands of years.'

"

My anvil is well over 100 years old, so carries the energy and history of the craft within it. It's also the one I learnt on. Along with my cross pein hammer, it's one of my favourite tools.

CARING FOR YOUR DAMASCUS STEEL KNIFE

A European-style chef's knife like this one is your kitchen workhorse, used for the majority of your cutting and slicing. Iain recommends using an end grain, soft wood timber chopping board – never glass. Never scrape the edge of the blade across the board, put it into a dishwasher or sink full of water, heat it on a hotplate or oven or use it as a lever or hammer.

High-carbon steel knives will mark or rust as they don't have chromium in them, like stainless steel does. This means the knife won't stay shiny, but it will stay sharp for longer. The blade will develop marks or a grey patina over time. This won't affect its integrity.

To prevent rusting, keep your knife as clean and dry as possible – don't wash it up, simply wipe it clean, and apply a light coating of oil after use. Olive, vegetable, canola and sunflower oil are all suitable for chef's knives. To create a water-repellent environment to store your knife in, pour some oil into your sheath or knife block then pour it out again. If you have a timber handle, rub some oil into that too, to keep it from drying out.

Your knife should only need to be put on a sharpening stone once a month and on a sharpening steel once a week, even if you are using it on a daily basis.

HOW TO

SHARPEN A KNIFE

MATERIALS

- Water

TOOLS

- Flat sharpening stone (1000+ grit)
- Rubber foot for stone or a folded damp tea towel
- Timber wedge cut to 15 degrees
- Honing steel

METHOD

1. Set yourself up for sharpening. You'll need a work surface that can get wet, such as the draining board of your sink. Place your sharpening stone on the rubber foot or a folded damp tea towel to hold it in place.
2. Wet the stone liberally with water.
3. Place the timber wedge on the stone; this will be your guide to achieving the correct angle of blade to stone.
4. Gently place the flat of the blade against the wedge and slide it down until the edge touches the stone.
5. Now, holding the handle firmly but flexibly, maintain the angle set by the wedge with one hand and use your other hand to remove the wedge.
6. With firm (not hard) pressure, make a slicing motion along the length of the stone, maintaining the angle all the while. Make sure to use the whole surface of the stone and the whole cutting edge of the blade in each stroke. For longer blades, hold the spine of the knife with your other hand, in a pincer hold, for control.
7. Repeat this action ten times on the same side of the blade, then switch sides and repeat. Inspect your blade edge by angling it towards a light source. Any shiny spots are in fact blunt areas of the edge – you need to continue sharpening until they're gone.
8. Once you're satisfied that there are no blunt areas, take your honing steel in your non-dominant hand, pointing away from you. Take your knife in your dominant hand and make a slicing motion on top of the steel, away from you, maintaining a 15-degree contact with the edge of the blade, down the length of the steel, ensuring the whole cutting edge of the blade makes contact in each stroke. For the opposite side of the blade it will be a slice along the underside of the steel. Do this ten times, alternating top and bottom each time, then inspect or test for sharpness.
9. Continue until you are happy with the sharpness. A sharp knife should be able to slice effortlessly without grabbing or catching, without requiring heavy pressure.
10. Wipe any residue from your knife, dry, then oil lightly. Your knife is now ready for use, or storage. Chop chop.

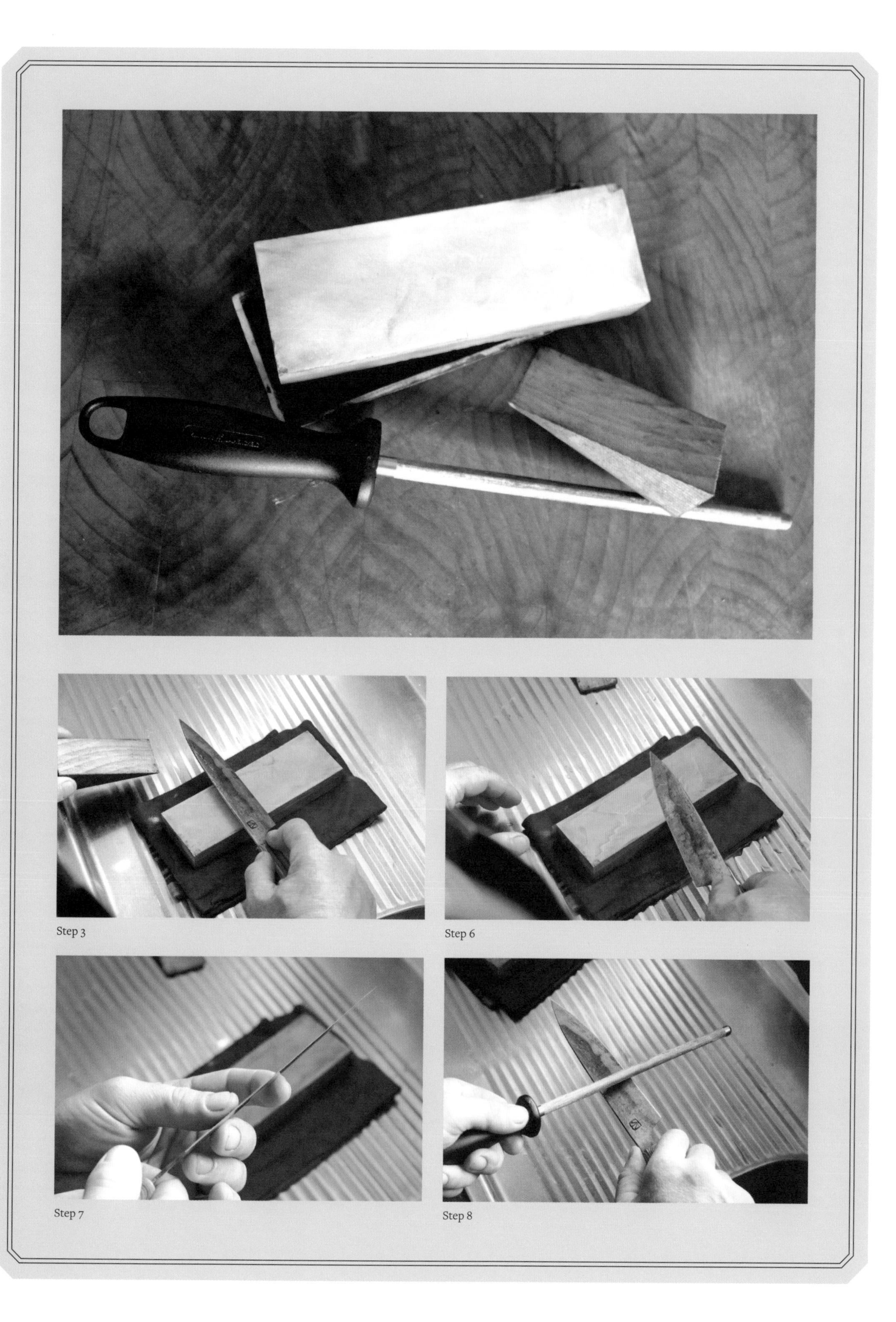

Step 3

Step 6

Step 7

Step 8

hapfo
hapfo

SPALTED BEECH BOWL

Woodturning

Woodturning is the craft of shaping wood on a woodturning lathe using hand tools. The trick with making bowls is to achieve a consistent shape, free of bumps or flaws; that's not as straightforward as it sounds. Franz Keilhofer, a part-time maths teacher, fashion model, farmer, and singer in a German hardcore band, has mastered the craft. He turned these bowls from spalted beech on his family's farm in Upper Bavaria, Germany. In nature, spalting occurs when green wood is left unprocessed for a long time. Natural decay starts and algae, fungi and sponges begin to grow through the log, creating striking patterns. Franz's spalted wood is created in a controlled, fungi-friendly environment to speed up the spalting process and make the resulting patterns a bit less random. These bowls are food safe, and are intended for fruit, salad, cereals and the like. Mmm, breakfast.

MEASUREMENTS	Diameter: 30 cm (12 in); height: 10 cm (4 in)
MATERIALS	Spalted beech, PVC glue, finishing oil
KEY TOOLS	Detail gouge, bowl gouge, branding iron, fine brass wire brush, fine sanding fleece
KEY MACHINES	Band saw, chainsaw, sandpaper, woodturning lathe
TIME TO MAKE	1 hour
LIFESPAN	Heirloom quality

FRANZ KEILHOFER — Woodturner [Upper Bavaria, Germany]

German woodturner Franz Keilhofer can turn around 160 bowl blanks (that's the roughly turned shape of a bowl) in a day. But figures like that are deceptive, as the turning itself is only a small part of Franz's process. It takes around a year and a half to take the wood from tree to finished bowl.

Franz lives in the Bavarian Alps, and sources and cuts the wood he uses himself.

'We are surrounded by mountains, small fields, forests and farms. We have our own small forest, and a lot of our neighbours have a lot of wood. I always try to get wood from trees that must be cut down for good reasons,' says Franz.

He transports logs to his farm with a tractor, then takes them into his workshop, an 80 sq m (861 sq ft) converted warehouse, where he uses a chainsaw to fillet the log into smaller pieces.

'The size I make the blanks depends on the size of the bowls I want to make. The wood tells me what needs to be done, and what its best purpose is. In an ideal world, there is only one perfect bowl in a piece of wood and that needs to be found.'

There are several aspects to consider when making a bowl that is to last for a very long time. 'It starts with the selection of the right log and the positioning of the bowl inside it, and ends with the right finish. Design also needs to be timeless. A good bowl will please everyone from the grandparents to the kids without being old fashioned, or fashionable.'

Once he has a finished bowl in mind, Franz uses the band saw to cut the blanks into a circular shape. He then mounts the bowl onto the lathe and uses a bowl gouge to roughly turn the outside, then flips it and roughly turns the inside.

Next he uses PVC glue to seal the end grain areas – the grain of wood seen when it is cut across the growth rings – to prevent the wood from cracking during the drying process.

'Now I leave the bowl in my wood store, a wooden shack with shelves stacked with bowls in different woods in various stages of

the drying process, for between six and twelve months, until the wood has dried completely. During this time the wood warps, shrinks and changes its form into an oval, elliptical shape.'

When the wood is dried, Franz mounts the bowl on the lathe to true (make round again) the outside and the inside. He shapes it to its final form and refines the surface with a delicate cut from a freshly sharpened bowl gouge.

Next, he uses sandpapers ranging from 80 to 1500 grit to smooth the surface of the bowl, and remove any tool marks and scratches. He starts with the coarser (lower) grits and works his way up to the fine ones as the sanding progresses. Sanding is the part of the process that frustrates Franz the most.

'It's very important, but boring and dusty too. I aspire to a very high surface quality, so I sometimes have to sand a piece a second and third time until my inner monk is pleased. I get through this with audio books. They really help.'

Once Franz is content with the surface of the bowl, he brushes the wood with a fine brass wire brush to highlight the grain and remove dust, and applies a food-safe finishing oil. He leaves the bowl in his bowl storage room to dry for several weeks, then polishes the surface with a fine sanding fleece.

He finishes the base of the bowl, then uses a hot iron to brand it with his maker's mark.

'My hands are my most important tools. A craftsman is nothing without his hands and the hands are the connection between the brain and every tool that is held in the hands.'

CARING FOR YOUR WOODEN BOWL

Never place a wooden bowl in a dishwasher or on a heater, or leave it in a high-moisture environment for a long time. Clean it with a soft dry towel, or a slightly wet towel if the task demands it. There is no need to refresh the surface, but if you use the bowl for food, its inside should be wiped with a generous amount of food-safe oil, such as olive, linseed or sunflower, before use for the first few weeks of use, to give it a protective coat. Store in a dry place out of direct sunlight between uses.

"

The wood tells me what needs to be done, and what its best purpose is. In an ideal world, there is only one perfect bowl in a piece of wood and that needs to be found.

HOW TO

TURN A WOODEN BOWL

MATERIALS

- Chunk of wood
- PVC glue
- Finishing oil

MACHINES

- Band saw
- Woodturning lathe
- Electric drill (for sanding; optional)

TOOLS

- Screws to fit the lathe's faceplate
- Bowl gouge
- Skew chisel
- Detail gouge
- Woodturning chuck
- Sandpaper (80–1500 grit)
- Fine brass wire brush
- Polishing fleece

SAFETY GEAR

- Safety goggles
- Earmuffs

METHOD

1. Hold the wood. Think about its size and shape and what kind of bowl might be within.
2. Don your protective gear – glasses or goggles and earmuffs – then get to work.
3. Use the band saw to cut the wood into a circular shape.
4. Use the screws to attach the faceplate to the wood.
5. Mount the bowl (with the help of the faceplate) onto the lathe.
6. Use a bowl gouge to roughly turn the outside.
7. Use the skew chisel or the detail gouge to cut a tenon on the exterior base of the bowl, so it can be mounted onto the chuck of your lathe.
8. Mount the bowl the opposite way to roughly turn the inside.
9. Seal the end grain areas with glue to prevent the wood from cracking during the drying process.
10. Leave the bowl to sit in a cool, dry place for up to a year. During this time, the wood will warp, shrink and change its form into an oval, elliptical shape.
11. Attach the bowl to the lathe with the help of the tailstock to true the outside and the tenon.
12. Shape it to the final form and refine the surface with a delicate cut from a freshly sharpened bowl gouge.
13. Now it's time to finish the inside. True the shape again, then refine the surface with a freshly sharpened bowl gouge.
14. Use the electric drill to smooth the surface using sandpapers ranging from 80 to 1500 grit. Begin with coarser (lower) grit and work your way up to finer ones. If you don't have an electric drill you can do this part by hand.
15. Brush the wood with a fine brass wire brush.
16. Apply finishing oil.
17. Polish the surface with a fleece after the oil has dried.
18. Finish off the bottom of the bowl by removing the tenon with a bowl gouge.

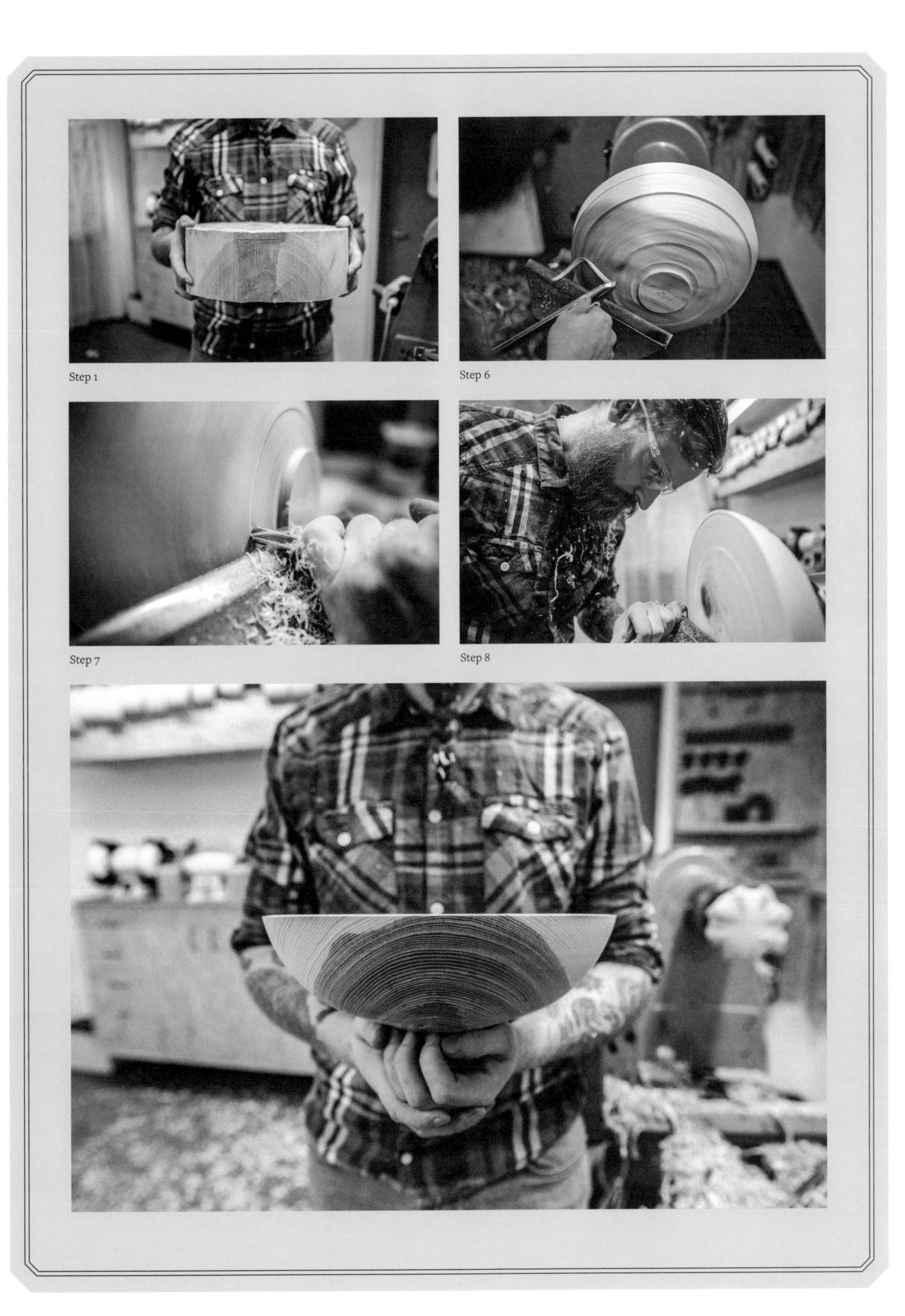

Step 1

Step 6

Step 7

Step 8

STONEGROUND WHOLEMEAL SOURDOUGH BREAD

Baking

People have been breaking bread together for thousands of years. Eating bread is a cultural and culinary experience, and gives us a way to digest grain as a food source. Wheat, the staple of bread as we know it, was first domesticated in the western half of Asia sometime around 8000 BCE. Cultivation soon spread to North Africa and on to Europe, and the ability to farm, process and cook with wheat is a bona fide hallmark of early domestic societies. Leaven, also called the starter or the mother, is the wild ferment that makes a loaf of bread rise, and is an ancient part of bread baking culture. Renowned Australian baker and sourdough advocate Ken Hercott has lived the whole bread story, from soil and seed to table and belly, and is passionate about introducing people to the pleasures of home baking.

MEASUREMENTS	Approximately 500 g (1 lb 2 oz) of wholemeal flour per 900 g (2 lb) loaf
MATERIALS	Freshly stoneground and organic wholemeal flour, leaven, oil, unrefined salt, water
KEY TOOLS	Glass jar, loaf tins or baking trays, mixing bowl, proofing baskets, wooden spoon
KEY MACHINES	Peel, wood oven
TIME TO MAKE	15 hours
LIFESPAN	Bread: one week, leaven: indefinite

KEN HERCOTT — Baker [Maldon, Australia]

Ken Hercott became interested in bread and baking at an early age; growing up on an organic wheat farm can do that to you. He taught himself to bake using wheat milled on the family farm, and sold his first loaf of bread to neighbours in the nearest town, Swan Hill, when he was just 19.

'My family went organic on the farm in 1983, when I was 13. We bought an old Yugoslavian stone mill and set it up on the farm; we milled our own wheat into flour for sale, and I started to bake with it,' recalls Ken.

'I first discovered sourdough when I came across the term "leaven" in some Christian literature. The word intrigued me. I did some research and learned that the roots of the word mean "to become sour", and that led me to sourdough.'

Ken's passion for bread and baking has taken him all over the world. In 1993, when he was 23, an informal apprenticeship in Massachusetts, USA under master baker Richard Bourdon of Berkshire Mountain Bakery raised Ken's baking skills to a professional level and cemented his commitment to producing quality bread.

Next came ten years of baking and honing his wood oven skills in Melbourne. Then to Europe with his family in tow, where Ken embarked on a Parisian bakery crawl, helped foster the reemergence of 'real bread' in Britain, and lived, baked and taught on an organic farm in Tuscany for two years.

For Ken, quality bread means a real food. Not processed white bread, but a slow stoneground meal, with character, crust and a flavour to match.

'We built a wood-fired oven in Tuscany, and I baked every Friday. We'd fire the oven for four hours, bake pizzas for lunch, then bake three loads of bread. The next day the oven still contained enough embedded energy for more delicate baking, like cakes and biscuits.

'Baking in a wood-fired oven is simple and sustainable; being on the land and baking bread made from flour milled from wheat from the very same land, with fuel from the land; it's a beautiful feeling. The bread tastes better than in a gas or electric oven – and the crust is superb.'

Ken and his family returned home to Australia in 2011, and today they call the historic gold rush town of Maldon, central Victoria, home.

'We moved here to make our own home, and to launch Bread Builders, my carpentry and backyard bread oven building business,' says Ken, who is also a ticketed carpenter.

He uses a starter that was gifted to him on his return by John Downes, one of the forefathers of the Australian sourdough movement. It's more than 40 years old, says Ken – though he cautions against getting caught up in the romanticism of the age of your starter.

'In Jewish culture, traditionally at Passover all leaven is purged from the house; it represents the Exodus. A new starter is created every year; a similar thing happens in Italy.'

And of course, he still tends his starter, and bakes bread once a month, for the local Maldon Market. He also teaches bread baking. Ken follows a French technique with his own leaven, which he believes improves the sweetness and texture of his bread.

'Feeding the leaven is a simple process of adding flour and water. I don't maintain a liquid starter, I keep a piece of dough aside as the mother for the next day's bake. I'm only baking once a month, so I keep it in the fridge and just take it out and start feeding it three days before I want to bake.'

'Coming from a farming background, I know the energy that goes into growing a cereal grain. When we make white flour, the energy we put into growing this incredible food source is processed out to the point where too much nutritional value is gone; we're just eating white starch. Whole grains hold incredible nutritional value; we need to eat more of them.'

> "Baking in a wood-fired oven is simple and sustainable; being on the land and baking bread made from flour milled from wheat from the very same land, with fuel from the land; it's a beautiful feeling.

CARING FOR YOUR LEAVEN

Leaven, also called the 'starter' or 'mother', is the wild ferment that makes a loaf of bread rise. Methods for maintaining leaven vary; Ken prefers the old French method of simply taking a piece of dough from the loaf of bread you're making and keeping it aside for the next one. This prevents over fermentation, which means your leaven won't (or shouldn't) develop an unpleasant, acidic flavour. It can be kept in a glass jar in the fridge.

When you want to make bread, remove the leaven from the fridge and feed it with flour and water in a ratio of 1:1 at twelve-hour intervals, at least three times before using it. This will reactivate it and bring the leaven up to the correct volume for the amount of bread you intend to make.

HOW TO

MAKE YOUR OWN LEAVEN

Most sourdough bakers will happily share some of their starter, and it's certainly easier to obtain and feed mature leaven than to make your own. However, meeting the challenge of making your own from scratch is rewarding. Everything has to start somewhere, right?

INGREDIENTS

- Stoneground wholemeal flour
- Water

TOOLS

- Large glass jar with lid
- Measuring cup

METHOD

1. Mix 1 cup of flour and 1 cup of water in the glass jar.
2. Once mixed, leave the jar covered out of the fridge (but not in direct sunlight).
3. After the first 24 hours, stir the mixture. If it is elastic, let it continue without adding anything and check it in another 24 hours. If it has lost its elasticity and has started to bubble and smell fermented, it is ready for feeding.
4. Feed it by composting half the amount, then adding another cup of flour and another cup of water to the remaining mixture in the jar. This should start to ferment within the same day, but do give the starter time to mature before the second feed. Many young starters fail because they are fed too soon, before they've fully fermented. A new leaven should be fed at least three times before it is mature enough to use in a dough.
5. When the leaven is ready for bread making, it is considered 'active'. It should look bubbly and aerated, and have a yeasty or beery smell. Leaven can be stored in the fridge until you are ready to use it. To prepare it for using, reactivate it by taking it out of the fridge until it returns to room temperature, then feeding it with flour and water in a ratio of 1:1, until you have enough leaven for the amount of bread you wish to make.

Step 1

Step 3

Step 5

HOW TO

MAKE WHOLEMEAL SOURDOUGH BREAD

Use your starter to make delicious fresh and nutritious bread. The process laid out below has several stages with rests in between – all up it will take 9 hours or so. It's a good idea to make the dough the night before you want to bake, so it can be left to prove overnight.

INGREDIENTS

- Stoneground wholemeal flour – preferably freshly ground and organic
- Water
- Unrefined salt
- Leaven
- Olive oil

TOOLS

- Measuring cups
- Mixing bowl
- Wooden mixing spoon (optional)
- Scales
- Clean tea towels
- 2 × 900 g (2 lb) loaf tins

PRO TIPS

- *The leaven content can be varied based on its activeness and your time schedule. More leaven will reduce the proofing time, but may also increase the sour flavour of the bread. An immature leaven will generally require a higher ratio as it is less active.*
- *The temperature and humidity of the baking room will affect the dough. For example, in hotter weather, use less leaven and chilled water.*

METHOD

1. Prepare your ingredients in the following ratios: 100% flour : 80% water : 2% salt : 20% leaven. One kg (2 lb 3 oz) of stoneground wholemeal flour should be enough for two 900 g (2 lb) loaves.
2. Combine the ingredients in the mixing bowl and mix them together until they are combined. Take care to prevent direct contact between the leaven and the salt.
3. Let the dough sit for 15–45 minutes, making sure the flour has absorbed the water before kneading. This is the start of the fermentation process.
4. Tip your dough out onto a benchtop and use a push and pull motion to stretch the dough and fold it back. Kneading develops your dough's elasticity and gluten structure. Knead it for 5 minutes, then let it rest for 5 minutes, continuing for a total of 30 minutes.
5. Now it's time for the first stage of proving (or bulk proof); leaving your dough for 1–8 hours to rise and develop its structure. Oil a bowl or tub, then place your dough in, making sure it's well covered to keep it from developing a skin.
6. Once the dough is proved, it needs to be divided into the size of a loaf. A 900 g (2 lb) loaf is a good size. Cut a piece off your dough and weigh it. Round it into a ball and sit it on a bench, under a clean tea towel. Leave it for at least 15 minutes.
7. Keep a small ball of dough, approximately 150 g (⅓ lb), aside as your leaven. Place it in a glass jar and put it in the fridge.
8. Oil your loaf tins.
9. Use your hands to give each loaf its final shape, then place them in the tins. Cover the dough with a damp tea towel and keep it in a humid environment, or cover it with a plastic bag, ensuring the plastic doesn't stick to the dough. Leave it for approximately 3–4 hours, but monitor the progress of the dough carefully. Check its consistency by pushing the dough gently with your finger. If it does not spring back, it is ready to bake.
10. Now it's time for baking. Preheat your oven to 240°C (465°F) then place the loaf tins in the middle of the oven.
11. A tinned loaf requires around 45 minutes to bake. Reduce the oven temperature to 200°C (400°F) for the final 15 minutes.
12. Check your bread to see if it is baked through by tapping it on the bottom. If it makes a hollow sound, like a drum, it's done.

Step 1

Step 2

Step 2

Step 9

iris hantverk:

iris hantverk:

SHOE BRUSH LOVISA

Brushbinding

Hot tip, shoe lovers. Never, ever use a synthetic brush to clean your leather uppers. It might be easy on your pocket, but it'll be hard on your shoes. When a synthetic brush is rubbed over leather, it charges the dust in the air and draws it into the bristles. This damages the bristles ... and it will damage your shoes, too. A synthetic brush will also need replacing sooner than you'd like. A good-quality shoe brush crafted from natural materials like the Shoe Brush Lovisa, which was handmade by visually impaired craftsmen at a small manufactory in Stockholm, Sweden, will last a lifetime. It's crafted from beechwood and dark horsehair that has been washed, combed, boiled, dried, and made into trusses. This brush is ideal for cleaning dirt off shoes before applying cream, and fits your hand like it was made for it.

MEASUREMENTS	Weight: 78 g (3 oz); length: 14 cm (5½ in); width: 5 cm (2 in); height: 4 cm (2 in)
MATERIALS	Beechwood; dark horse hair; stainless steel wire
KEY TOOLS	Crochet hook attached to a finger ring
KEY MACHINES	Cutting machine; drill press; manual portioning machine; sawing, milling and sanding machine; steel-cutting machine
TIME TO MAKE	15 minutes
LIFESPAN	20 years +

ÅKE FALK — Brushbinder [Stockholm, Sweden]

Åke Falk has been making brushes at Iris Hantverk since 1990. He learned his craft over two and a half years as an apprentice to a brushbinder called Gunnar Holmer; Gunnar himself learned his trade from another of Iris Hantverk's brushbinders before him. The Shoe Brush Lovisa is one of Åke's favourite brushes to make.

'I like making the shoe brush because it's tricky to make and I have to think about what I'm doing to make it right, even though I've done it a thousand times,' says Åke.

There are no shortcuts to making an excellent brush. A local woodworker prepares the wood for the handle before it arrives at Åke's workstation. The woodworker measures, saws, mills and sands it, then it's drilled with the number of holes required for the bristles. Every brush is different – some brushes, like the computer cleaning brush, have just 14 holes, while others, like a tailor's clothes brush, have 255. This shoe brush has 86 holes.

After the handle has been oiled with boiled linseed oil it's time for Åke to bind the brush. This means attaching each individual bundle of bristles to the wooden handle. He takes a length of fine stainless steel wire, threads it through a crochet hook that is attached to a ring he wears around his little finger, and pulls it up in a loop through the first of the smaller holes in the wood. He takes a bundle of bristles from the portioning machine in front of him (which sorts the horsehair into perfectly even bundles), places them in the wire loop, then pulls it up, into and through the other, larger side of the hole. The wire is looped, threaded and pulled tight on the top side of the wood, then the threading of the next hole begins. Åke works his way methodically around the brush in this fashion, tying it off at the end by pulling the wire through the hole, gently lifting one loop and pulling the wire through it twice. Finally, the ends of the bristles are cut and hackled (raised) on a cutting machine to make them smooth and even.

'I work quickly and methodically but I´m also very careful. It's important that the bundles are carefully separated and are consistent. My work must always be of the highest quality; at

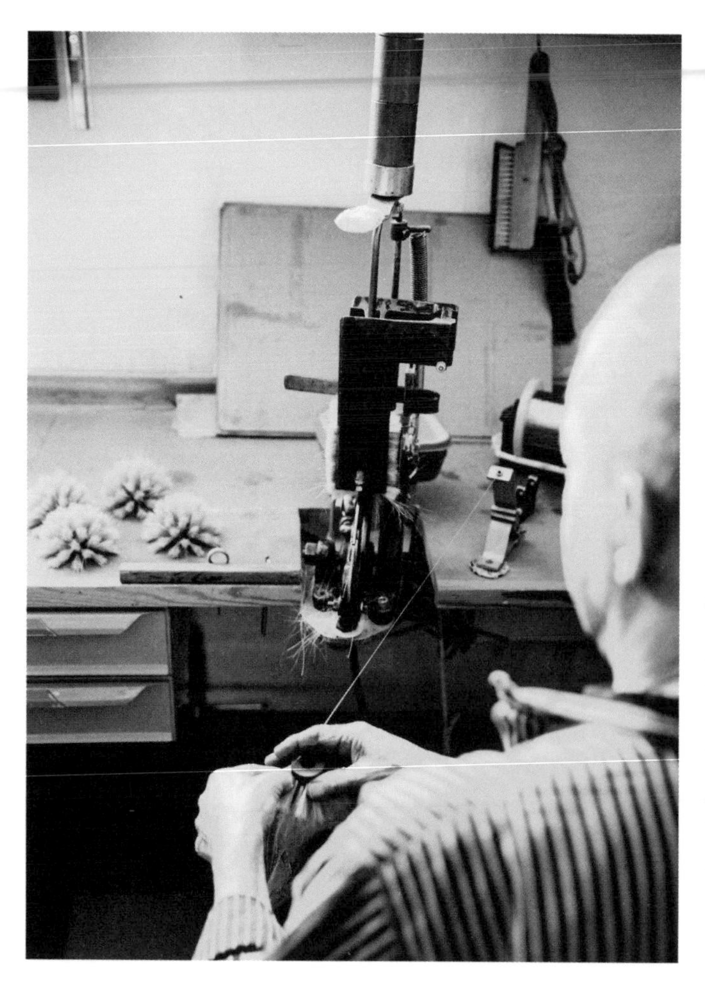

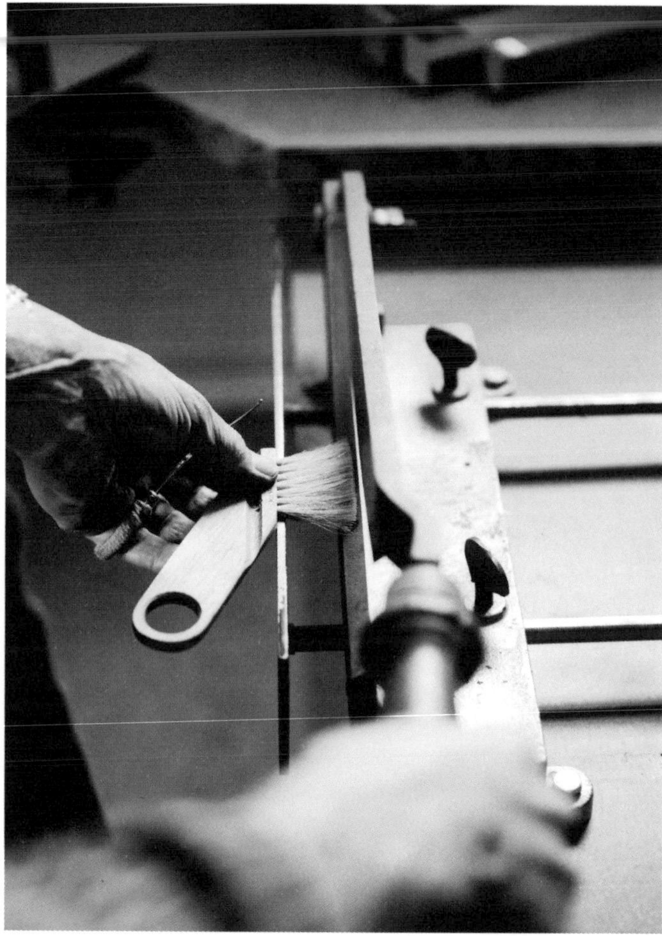

Iris Hantverk, I feel like I'm a part of a team of great craftsmen.'

The current owners of Iris Hantverk, Richard Sparrenhök and Sara Edhäll, bought the company in August 2012, when Sweden's *Socialdepartementet*, or Department of Social Affairs, withdrew the funding that had been the manufactory's mainstay since the 1950s. Together they set about rejuvenating the business with new designs, while retaining a tradition of care and respect for its workers.

'It's an old tradition in Sweden that visually impaired craftsmen bind brushes. Because their sight is impaired, the importance of other senses like listening and touch are reinforced in the craftsmen. They develop a real feel for the craft. It makes no difference to the quality of the brush whether a visually impaired craftsman binds the bristles or not, but for us, it's very important to preserve this profession for visually impaired people today,' says Sara.

Iris Hantverk's history stretches back to 1870, when a workhouse for visually impaired craftsmen was formed. Now, as then, every Iris Hantverk brush is hand bound by visually impaired craftsmen. A team of five workers produces around 100 types of brushes, from tailoring and shoe brushes to dishwashing and body scrubbing brushes. The intimate relationship they develop with their tools brings a whole new meaning to the notion of something being made by hand.

CARING FOR YOUR BRUSH

With a little simple care, an Iris Hantverk brush will age beautifully. Clean the brush with washing-up liquid and warm water. Let it dry, bristles down, on a clean towel, and oil the wood with food grade oil, or boiled cold-pressed linseed oil, from time to time.

I like making the shoe brush because it's tricky to make and I have to think about what I'm doing to make it right, even though I've done it a thousand times.

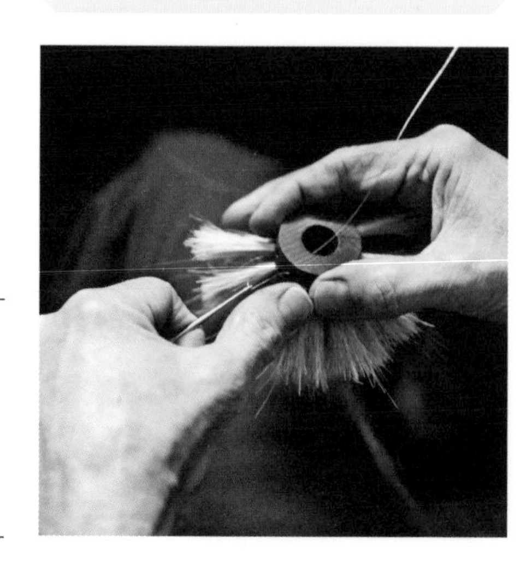

HOW TO

MAKE A BRUSH

MATERIALS

- A block of wood – oak or beechwood is best
- Boiled linseed oil
- Bristles – bundles of horsehair if you can get it, or a sturdy natural fibre such as grass
- Thin stainless steel wire

TOOLS

- Pencil
- Vice
- Hacksaw
- Sandpaper
- Hand drill or electric drill
- Clean, dry cloth
- Sharp scissors or cutting machine

METHOD

1. Design your brush. Keep it simple, and think about how the brush will fit in your hand. Are you left- or right-handed? Do you have long or short fingers?
2. Mark up the shape on your piece of wood, then place it in the vice and cut out the shape with a hacksaw.
3. Sand the wood until it's smooth.
4. Take your pencil and mark the spots for the holes on the underside. Put the brush back in the vice, then use your drill to carefully make a hole where each of the marks is.
5. Sand the wood again, if necessary, then oil it by applying boiled linseed oil to a clean, dry cloth and gently rubbing it into the wood. Leave it to dry overnight.
6. When the brush is dry, it's time to attach each individual bundle of bristles to the base of the brush. Loop the stainless steel wire around the first bundle of bristles, then thread the wire through the first hole.
7. Pull the wire through the hole and secure it by forming another loop around the bristles, then twist the wire around itself twice.
8. Repeat this step at the next hole, continuing your way around your brush until you are back at the first hole.
9. Tie off the stainless steel wire by twisting it around; it should snap easily.
10. Trim the ends of the bristles with a pair of sharp scissors (or a cutting machine, if you have one) to ensure they're even.

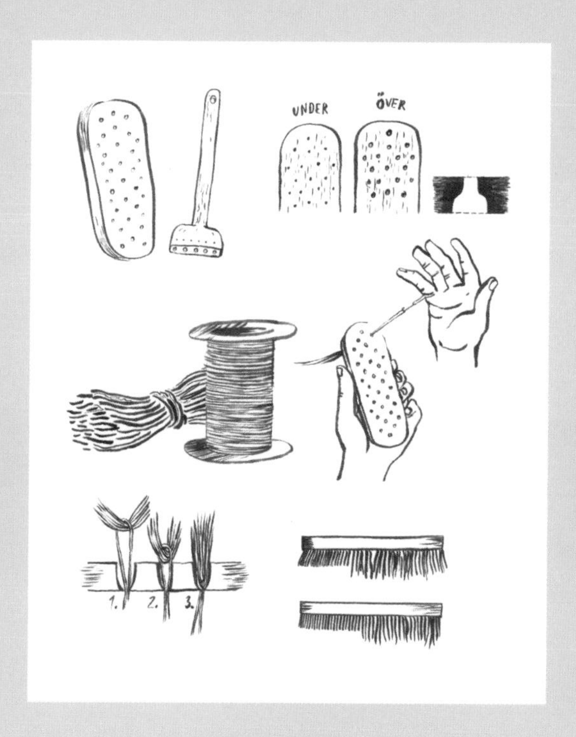

Di
Mi
2
5
3
6
4
7
8
5
9
6
10

ETERNAL CALENDAR

Woodworking

Humans have been keeping track of time since, well, the beginning of time. Our current approach to the turning of the planets, seasons, and the advent of night and day is shaped by the Gregorian calendar, which was introduced by Pope Gregory XIII in October 1582. Defining time is one thing; keeping track of it is quite another. This clever calendar was made by a Japanese-born, Berlin-dwelling designer and maker. It looks and works a bit like a wall clock, and will last for as long as our calendar system does. However, it's not battery powered – it's human powered ... The idea is that every day, you'll manually move the hands to their appropriate spot, adjusting the date and day of the week, and of course as the months roll over, you'll shift them along too. Talk about analogue.

MEASUREMENTS	Diameter: 31 cm (12 in); depth: 19 mm (¾ in)
MATERIALS	Beechwood, glue, ink, paint, paper, spruce, threaded rod, varnish, wooden screw nut
KEY TOOLS	Japanese scriber gauge
KEY MACHINES	Drill, milling machine, router, sander, spray gun, suction, table saw, thickness plane
TIME TO MAKE	6 hours
LIFESPAN	Until the end of time

SHIGEKI YAMAMOTO — Furniture maker [Berlin, Germany]

The most important tool in Shigeki Yamamoto's toolkit is a scriber gauge. It was gifted to him in 2001 by his then-boss, a Japanese iron sculptor by the name of Nobuyuki Tachibana. The scriber gauge enables Shigeki to accurately mark distances from and along wooden edges, and is invaluable in his furniture making and design.

Of course, Shigeki took the scriber gauge with him when he relocated from Osaka in Japan to Berlin in 2006.

'I was keen to be closer to European design culture. Since coming to Berlin I've trained as a carpenter, and now I focus all my work on wooden furniture and forms,' says Shigeki.

Shigeki developed the concept for the eternal calendar in 2011, after watching the hands of a wall clock ticking around and around and making the link that days and months also move cyclically, but that we don't generally mark them in the same circular way.

He set about working out how to mark all these measures of time on the same timepiece.

'I did a lot of prototyping. There were around six versions – it was particularly difficult to find the correct distances and positions for the numbers and names of the month.'

To start making the calendar, Shigeki mills the beechwood that forms the calendar's base into a round shape, using a router.

'Next I use a sander to mill the edges until they're smooth, then smooth all the surfaces with sandpaper. I brush the edges twice with the milling machine, to give the wood a softer appearance.'

Shigeki flips the base over and drills a hole for the hook in the back, then oils the beechwood. Next, he grinds the hands, and drills holes in them. Then he spray-paints all the components: blue for the days of the week; yellow for the name of the month and red for the days of the month.

The paper that forms the calendar's face has already been cut into a circular shape measuring exactly 31 cm (12 in) in diameter and printed with high-grade ink onto high-grade paper by an offsite printer. Shigeki carefully glues the paper onto the beechwood.

'This part is quite tricky,' says Shigeki. 'I need to apply the glue as uniformly as possible, then apply as little pressure as possible when applying the paper to the wood, lest it mark.'

To finish the calendar, he drills a hole in the centre of the face and inserts the wooden screw nut, then the hands, which he also crafts from wood.

Shigeki makes around a dozen calendars at a time in a 32 sq m (344.4 sq ft) workshop in a former factory space just south of the former Tempelhof Airport in Berlin. The entire space has been converted to artists' studios; Shigeki's, on the third floor, is one of more than 30.

Shigeki was born and raised in the Osaka prefecture of Japan. His father was an engineer at a large industrial plant and a maker of things, and he had a big influence.

'When I was six years old my father built a bed for me in the top of a typical Japanese wardrobe. It was made from wood, and I needed a small ladder to climb up and get into it. I loved it.'

It's easy to see the link between a bed like that and the playfulness that is a deliberate feature of his designs. Nobuyuki Tachibana was another big influence. Working with the renowned iron sculptor was the first job Shigeki had after graduating from his studies in interior design at the Osaka Sogo College of Design. One particular item Shigeki made, a metal chair, led him to consider making his career in furniture.

'When I launched my brand in 2010, the leading concept was "High Humour Design". I like to bring a sense of fun into daily life, while maintaining a commitment to quality. I also strive to infuse my work with spiritual richness in the face of today's mass production.'

"

I like to bring a sense of fun into daily life, while maintaining a commitment to quality. I also strive to infuse my work with spiritual richness in the face of today's mass production.

CARING FOR YOUR ETERNAL CALENDAR

Hang the calendar out of direct sunlight, and dust it occasionally. Always make sure you have clean fingers when you shift the hands along.

HOW TO

MAKE A SHIGEKI YAMAMOTO FLOWER VASE

MATERIALS

- Test tube
- A piece of beechwood or other wood of your choice measuring approximately 7 × 7 cm (2.8 × 2.8 in) - the length must be slightly longer than the test tube
- Decor wax

TOOLS

- Pencil
- Hand saw or circular saw
- Measuring tape or ruler
- Drilling machine or hand drill
- Router
- 240 grit sandpaper
- Brush or cloth
- Cloth
- Stitching pony or vice (optional)
- 2 Leather needles

METHOD

1. Align the test tube lengthways along the piece of wood, then use the pencil to mark the length of the test tube on the wood.
2. Saw through the piece of wood at the marked position using a hand saw or circular saw.
3. Take the pencil and mark the centre of each end.
4. Using the measuring tape or ruler, measure the diameter of the test tube and add 2 mm (1/10 in) to the figure.
5. Next, use the drilling machine or hand drill to drill a hole within and to the edges of the marked area. The hole should be the same diameter as the measurement you took in step 4.
6. Trim the edges at the top using the router.
7. Sand the surface with sandpaper.
8. Apply decor wax with a brush or cloth.
9. Insert the test tube into the drill hole.
10. Insert a flower and present the vase to the one you love. Or just put it somewhere nice.

Step 1

Step 5

Step 8

Step 9

Step 10

BLACKWOOD ROD BACK SETTEE

Chair making

Windsor chairs, with their solid, saddle-shaped seats, round-tenoned spindles and steam-bent backs, are one of the most comfortable – and oldest – chair designs known. Their exact origins are uncertain, but they were being made in England in the 17th century, initially by wheelwrights, who use similar tools and methods. This one was meticulously crafted in Kyneton, a town 83 km (52 mi) from Melbourne in Australia, by chair maker and woodworker Glen Rundell, who has spearheaded Australia's romance with old-fashioned trades through an annual Lost Trades Fair. This settee is made from Australian blackwood, which Glen selected, dried and crafted into its current form himself. It's a chair that will outlive you, and your children, and your children's children, and its design pedigree makes it a family heirloom from the outset.

MEASUREMENTS	Length: 120 cm (47 in); width: 50 cm (20 in); height: 105 cm (41 in)
MATERIALS	Blackwood, natural oil, wax
KEY TOOLS	Adze, chisels, drawknives, jointing plane, reaming tool, scorp, scraper, spokeshave, travishers
KEY MACHINES	Band saw, dimensioning table saw, handheld drill
TIME TO MAKE	4 days +
LIFESPAN	Heirloom quality

GLEN RUNDELL — Chair maker [Kyneton, Australia]

Glen Rundell hasn't been a chair maker all his life. At 16 he was an apprentice plumber, then a farmer, then he became a policeman. Nearly 20 years on, he has thrice travelled to the United States and studied chair making. Today, he makes and sells Windsor chairs under the name Rundell & Rundell, and teaches the time-honoured art of chair making.

Most Windsor chairs produced in Glen's workshop are made from a green log in his yard, or from a tree he salvaged or harvested locally. This means that before he can begin making an actual chair, he must prepare and steam or dry the wood. This alone can take one year, or more.

The seat stock, legs, stretchers and other turned components, Glen simply cuts and sets aside to air dry. Stock for spindles and parts intended for steam bending are split or sawn in the yard, then brought into Glen's workshop for dimensioning, steaming, then drying.

'My workshop is a ramshackle old building, which has been poorly extended, time and again, for about 90 years. The roof leaks, the central box gutter often overflows and out the back it's not much more than a hot tin shed in summer and an icebox in winter. But it's ours and it gets the job done,' says Glen.

A year or two down the track, when the wood is ready to use, Glen selects the best of the long boards for the seat.

'I'm looking for timber free of blemishes, knots or inclusions. Quarter sawing usually yields the best grain, visually, and is also easiest to carve.'

Glen thicknesses these rough sawn boards to approximately 5 cm (2 in) thick. He 'joints' both edges with a jointing plane by hand, places the boards edge to edge to check for a gapless join, then glues and clamps them overnight.

The next day, he transfers the seat pattern onto the seat stock and marks and drills holes for mortices – six legs, two arm posts, two back posts and 18 spindles – into the seat. He uses a reaming tool to taper all the mortices except the spindles and stretchers, then clamps the seat to the bench and sets about hewing and shaping it with various handheld, edged tools: first a chair adze, then a scorp, then travishers. He shapes the outer edges and rear of the seat, then uses a scraper blade to refine the final shape and surface quality of the whole seat.

With the seat done, Glen moves on to making the arm rests and back posts, then the crest rail and spindles, then the legs and stretchers. It's careful, delicate work, where he roughly shapes each piece of wood then moves in to give each its final form with the right tools for the job: variously drawknives, spokeshaves and scrapers.

Then he must ensure that the parts all fit perfectly together. Glen must get the alignment and angle of each piece right with its immediate neighbours, and then with every other piece in the settee. This means that the angle of each mortice and tenon must be exactly right, and it may take hours of measuring, sighting and checking to ensure each piece will give a precise and accurate fit.

When he is happy that all of the parts will fit precisely, he scrapes and sands the settee, ready for gluing. Using natural hide glue, Glen glues and wedges in the order of stretchers, legs, undercarriage, seat, back and arm posts, spindles.

For the final stages, he trims the bottoms of the legs so that they sit flat on the floor, and chamfers the edges of the legs. He scrapes and sands the settee, hand carves the false mitres at the joint of the crest to the back posts into a 'duck's bill' shape, trims the tops of the legs and arm posts, and scrapes all the surfaces again. He applies six coats of a natural oil and wax-based finish, rubbing each coat back with a superfine abrasive before applying the next coat. And then it's finished. It is a long and meticulous process, and one that Glen takes great joy in.

'Skills and craftsmanship take time, patience, lots of repetition and most of all dedication. There's so much to be learned from our past masters; we just have to open our ears and listen.'

“

Skills and craftsmanship take time, patience, lots of repetition and most of all dedication. There's so much to be learned from our past masters; we just have to open our ears and listen.

CARING FOR YOUR BLACKWOOD ROD BACK SETTEE

Fine furniture needs a custodian who will look after it until it is time to hand it down to the next generation. Don't drop it, don't use it as a ladder. Don't sit on the arm of it. Don't sit it too close to the fire. Don't sit it directly over or under the air-conditioning or central heating duct. Try not to leave it in full direct sunlight every day. Don't leave it outside in the rain and wind. Sit on it often, keep it clean. No need to oil the wood; allow it to age well and it will develop a beautiful, natural patina.

HOW TO

USE A TRAVISHER TO HOLLOW A SEAT

A travisher is a specialist chair maker's tool that is used to hollow and shape the concave curves of a Windsor chair seat. It's extremely responsive to its user; apply light pressure and it will cut so finely that the wood shavings will be translucent; apply more pressure and it will remove considerable timber with each pass. Use it to refine the seat's shape and surface quality after you've roughed out the shape with a scorp or adze.

MATERIALS

— Wood for Windsor chair seat

TOOLS

— Travisher

METHOD

1. Rest the wood for your seat on a stable work surface at approximately waist height.
2. Take the travisher and rest the base on the surface of the timber, at the edge closest to your body.
3. Now tilt it back until the blade edge makes contact with the wood's surface, then push it forward at a skewed angle to slice and shave the wood surface. It's important to work in one direction only – always away from yourself – to maintain the smoothness of the grain. Glen recommends starting with light pressure until you get the hang of handling the travisher.
4. Methodically repeat this motion across the surface, most often across or at an angle to the long grain direction of the timber, to create uniform, even patination.
5. Shave the entire concave surface of the seat in this way until you've achieved the desired depth, shape and surface quality.

Step 2

Steps 3–5

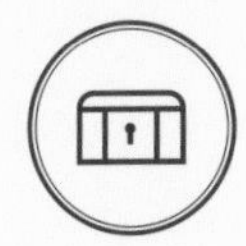

CAMPAIGN CHEST

Furniture making

When the first Duke of Wellington defeated Napoleon at the Battle of Waterloo in 1815, he had a campaign chest or two in tow. Named for the British military campaigns it was designed for, campaign furniture includes tables, chairs, beds, drawers and chests. Simply designed and sturdily made, it had to be strong enough to survive days and months on the road. This chest is intended for blankets, and was made in 2012 from incredibly dense and hard antiqued sugar maple by a studio furniture maker in Virginia, USA. Its design cleverly deviates from the classic campaign chest to create an even sturdier piece. The frame-and-panel construction on top and continuously dovetailed case enable the whole piece to expand and contract comfortably through the seasons, and mean cracks in the wood are very unlikely. Onward: march!

MEASUREMENTS	Width: 81.3 cm (32 in); depth: 52 cm (20 in); height: 58.4 cm (23 in)
MATERIALS	Ammonia, antiqued maple, brass, gel stain, hide glue, paste wax, shellac
KEY TOOLS	Bar clamps, dovetail chisels, dovetail saw, hand planes, hand router, steel wool
KEY MACHINES	jointer, planer, band saw, hollow chisel morticer, router table, table saw
TIME TO MAKE	100 hours
LIFESPAN	Heirloom quality

JEREMY ZIETZ — Furniture maker [Virginia, USA]

Ask heirloom furniture maker Jeremy Zietz about his most important tools and he'll admit to having a soft spot for his spokeshave ... but he refuses to play favourites. Jeremy considers all of his gear, from hand planes and protractors to bins of scrap blocks, vital to his process, and therefore equally valuable.

Jeremy's tools are a mixture of old and new that he has collected over time, ever since he began studying fine furniture making at the Vermont Woodworking School in 2011.

He shares core machinery, such as table saws, jointers and band saws, with a collective of makers working out of a 1524 sq m (16,400 sq ft) former elevator repair workshop on the north side of Richmond, Virginia.

The design of this campaign chest is clean and simple, but don't let that deceive you; it takes Jeremy around 100 hours to complete.

He rough mills the wood, lets it rest, then uses a jointer and planer to bring the wood to its final dimensions. He glues the edges to form the front, back and side panels of the chest.

Then he uses a table saw to rip (cut parallel to the grain) and cut out the drawer areas in the front panel, carefully putting the wood aside for later, to ensure a continuous grain pattern on the face of the chest.

Next, Jeremy gets started on the dovetail joints. This chest has 40 dovetails – 20 on the front and 20 on the back – beautiful and delicate joints whose name alludes to the shape of the joint. If done correctly, they'll give the chest immense strength.

'The dovetailing process is immensely rewarding. Cutting joinery like this by hand, to me, is the most direct way of joining wood together,' says Jeremy.

Now he uses a table saw to cut dado grooves for the chest's bottom panel on the inside of all four box panels, and then grooves for the 'web' frames that support the drawers. He mills, glues and surfaces the bottom panel with a smoothing plane.

After a dry fitting, Jeremy glues the case using hide glue, clamps it with parallel bar clamps, then glues up the big box, making sure to trap the bottom panel in its groove.

He makes the chest's top, and top panel, then cuts through the entire chest with a table saw.

'This leaves me with the main carcass and the lid. I make a mortice and tenon web frame to support the drawers and glue them into place, flush with the bottom of the drawer cavity. I set and glue guide rails for the drawers into place, then I make the drawers from the wood I saved earlier.'

There are more dovetails to carefully hand cut and shape here; 16 for each drawer.

Afterwards, Jeremy cuts relief scoops on the chest's bottom edges using a template and hand router, then cuts the relief mortices for inlaying eight brass corner braces and two drawer pulls. These are signature to the campaign style.

It's time to finish the chest. Jeremy mixes and applies aniline dye to the chest to antique the maple. Then a shellac wash coat, then a glazing gel stain, then more than 20 coats of hand-rubbed shellac.

'I shellac the chest using a French polishing technique, which gives a glossy surface. I bring this outer sheen to satin with steel wool and paste wax. I apply three coats of shellac to the inside of the chest, and none to the inside of the drawers.'

He waxes the exterior, and the drawer guides and runners. Then, he fumes the brass hardware in an enclosed bucket using ammonia, until a patina develops. He burnishes the chest with steel wool and installs the hardware: brass hinges and brass stays for the lid. The last step is to carve the date and his maker's mark into the base.

'Designing and making furniture is incredibly dynamic, which is what I love about being in my workshop. It is physical, creative, cathartic, and can be intense at times too. In the end, it's gratifying to remember the many generations each piece will live with.'

CARING FOR YOUR CAMPAIGN CHEST

After many years, the chest may need new finish applied to maintain its protective qualities. If any wax remains, remove it with some mineral spirits on a cloth. Then apply new shellac mixed with denatured alcohol. Otherwise, the chest will need no extra care.

"Designing and making furniture is incredibly dynamic, which is what I love about being in my workshop. It is physical, creative, cathartic, and can be intense at times too.

LIE-NIELSEN . USA

HOW TO

MAKE A DOVETAIL JOINT

Renowned for its superlative tensile strength – aka resistance to being pulled apart – the prettily named dovetail joint is commonly used to join two pieces of wood at right angles. A series of cleverly cut trapezoidal pins extend from the end of one board to interlock with a series of trapezoidal tails cut into the end of another board.

MATERIALS

- Wood to be joined – two sides of equal width and thickness
- Wood glue

TOOLS

- Cutting gauge
- Sharp pencil
- Dovetail square or sliding T-bevel gauge
- Vice
- Dovetail saw
- Coping or jeweller's saw
- Single bevel bench chisels
- Mallet
- Workbench clamp
- Plane or sander

METHOD

1. Set the cutting gauge's depth to be slightly thicker than the thickness of the sides of the wood. Now, with the fence of the gauge against the ends, score 360 degrees around both ends.
2. Use a sharp pencil to mark the dovetail layout using a dovetail square or sliding T-bevel gauge. Mark the pins (inverse from dovetails) with an 'X' to remind yourself of the waste portion, and ensure there are pins on each end of the joint.
3. Place the wood in the vice.
4. Take the dovetail saw and carefully make square cuts down the waste side of the dovetail marks, stopping at the depth mark.
5. Now cut out the bulk waste by removing the waste that you marked with the 'X' on each end, cutting slightly away from the depth line.
6. Use the coping or jeweller's saw to remove the middle waste.
7. Using very sharp chisels, set the chisel edge (bevel side out) in the scoring mark and cut square and inwards from the edges, using a mallet to gradually pare away the wood. This develops the 'shoulder' of the joint. Take care to ensure the shoulder and tail sides are square.
8. Using the dovetails as a template, align the shoulder and edge of the pin board. Now take your pencil and carefully trace the dovetails' shape onto the end of the opposing side. Mark and 'X' the waste portion, then, using the dovetail square, mark vertical lines down to the desired depth score mark.
9. Cut the dovetail shape and remove the outside and middle waste, as on the first side. If necessary, use the saw or chisel and mallet to carefully pare away any excess wood. Ensure the edges of the shoulder are square.
10. Do a test fit to ensure the joint will go together without excessive force. Make any adjustments necessary.
11. To permanently assemble, apply glue only to the sides of dovetails and corresponding pin sides and clamp the side with tails until the shoulder gap is closed.
12. After the glue is cured, plane or sand any slightly protruding dovetails and pins.

Step 1

Step 2

Step 3

Step 4

Step 5

Step 6

Step 8

Step 10

ThinkPad

NOVENA HEIRLOOM LAPTOP

Small-scale manufacturing

A computer, made to last? Wait, don't those things have planned obsolescence literally wired into them? Not this one. This is the Novena Heirloom, one of just a dozen laptops made by veteran camera designer Kurt Mottweiler and hacker and open-source champion Andrew 'bunnie' Huang, who is renowned for being the first to hack the original Xbox. Intended for serious tinkerers who want to know their computer from the inside out, it was conceived as a 'hacker's dream laptop'. Like its big brother, the aluminium-cased Novena, the clamshell style Heirloom is built with completely open source and modifiable hardware. It runs on open source Debian Linux operating systems, is powered by an ARM processor, and has a removable keypad that enables you to access the electronic hardware beneath. The Heirloom's case is handcrafted from wood and aluminium and even shipped with Allen keys.

MEASUREMENTS	Height: 4.9 cm (2 in); length: 32.4 cm (13 in); width: 24 cm (9½ in); case base weight: 816.5 g (1 lb 13 oz); weight when fully assembled with all electronic components and battery: 2.2 kg (4 lb 14 oz)
MATERIALS	Novena composite (Makore, Walnut, Black Limba and Afrormosia timbers for the solid wood and veneer stock, e-glass cloth, cork and epoxy), 6061 aluminium
KEY HAND TOOLS	Callipers, chisels, files, handsaws, micrometers, mallets, marking knives, punches, rasps, rulers, scrapers
KEY MACHINES	Bandsaw, clamping presses, CNC milling machine, CNC router, high-speed micro drill press, hydraulic presses, joiner, lathes, MacBook Pro with Fusion360 CAD-CAM and other software, planer, tablesaw, universal milling machine, vacuum degassing chamber, vacuum and pressure lamination equipment
TIME TO MAKE	24 months (including design and prototyping)
LIFESPAN	Heirloom quality

KURT MOTTWEILER AND BUNNIE HUANG — Designer/craftsman and hacker [Portland, USA and Singapore]

When Andrew 'bunnie' Huang was seeking someone to help bring the Heirloom version of his hacker laptop to life, Portland's Kurt Mottweiler was the obvious choice. The renowned maker has been drawing on his woodworking, metalworking and luthiery skills to fuse craft and technology for decades. Kurt mostly makes furniture and bespoke cameras, so a laptop case was a fine challenge.

'The overall Novena project bears my name, but Kurt's signature is emblazoned on every Heirloom. He's the hands and heart who put the Heirloom case together,' says bunnie.

Balancing the large scale of the challenge with the small scale of production defined Kurt's approach to the project. Even though he set out to create a design that could be reproduced, each of the computers was essentially a one-off, custom project.

Kurt's first big challenge was working out how to make a case that was strong enough to contain 1.4 kg (3 lb 2 oz) of hardware and withstand daily use, yet flexible enough to be handled time and time again, while staying true to the brief's curvaceous Heirloom aesthetic.

Kurt decided to craft the Heirloom's case from metal and wood, paying homage to the vintage hi-fi equipment of designers like Dieter Rams. Aluminium was used for the side panels of the enclosure, and a wood composite for the main panels – but not just any old wood composite. Drawing on his experience using cross-banded wood laminates for his camera designs, Kurt embarked on a series of trials to create an entirely new material. He settled on a combination of Makore, Walnut, Black Limba and Afrormosia

CARING FOR YOUR NOVENA HEIRLOOM LAPTOP

Treat the Heirloom like any finely made wooden object. Keep it in a soft case and handle it gently to avoid scratches and dings. Clean the wooden surfaces with a slightly damp cloth or a fine furniture wax. Clean the aluminium surfaces in a similar fashion, but with window cleaner, or a cloth dampened with a mild soap and water solution.

timbers for the wood veneer elements of the finished Novena composite.

The composite was the perfect material for crafting the wave shape that undulates across the width of the Heirloom's bottom panel. The technique is similar to the one used to craft the famous aerodynamic curves of World War II British Mosquito bombers.

'The Mosquito bombers were made with cross-laminated veneer layers. The result is similar to plywood, except that it takes the shape of a mould over which the forms are made. With the Heirloom, I varied the technique by replacing the inner layers of veneer with layers of fibreglass cloth and cork, bonded together with high-modulus epoxy,' says Kurt.

He went with the wave shape for the bottom panel because of its visual appeal, but also because of its ability to improve the strength and heat dissipation of the case. He says that developing and implementing the composite used to make the main parts of the enclosure was his favourite part of the creative process – that, and collaborating with bunnie.

'Trading thoughts and ideas about how things would be done was seriously enjoyable,' he recalls. Mostly the pair communicated long distance, from their studios in Singapore and Portland, but they did spend an intensive week together in March 2015, in Kurt's 111.5 sq m (1200 sq ft) Portland studio, part of a former furniture factory.

Of course, there were dozens of design challenges to work through, such as how to fabricate the rear panel of the Heirloom. The two composite panels that make up the bottom of the enclosure and the flat panel behind the LCD were fabricated using a vacuum bagging technique common in both wood veneer and high-tech composite fabrication processes.

'But in the case of the rear panel, which carries the speakers and switches and supports the back of the keyboard, the bend radius was tighter than the vacuum technique could reliably manage,' explains Kurt. His answer? A two-part hard mould that used woodworking clamps to hold the composite materials in the required shape until cured.

Other, smaller details, like the lidded box-inspired visible friction hinges – an aluminium plate that connects the interior's centripetal CPU cooling fan with the heat sinks, and a veneered hinge plate that connects the LCD panel to the main enclosure – were custom designed and made, and took months to finalise.

It's just the kind of technical, mechanical problem solving that Kurt, whose tagline is 'artisanal craft in the digital age', thrives on.

'Kurt achieved everything we set out to do. He created a timeless, striking piece of art, which hits a perfect balance between function and form,' says bunnie.

'I'm looking forward to seeing how the wood case ages and wears. I'm deliberately not handling my Heirloom with white kid gloves, so that it gets some character.'

“

Kurt achieved everything we set out to do. He created a timeless, striking piece of art, which hits a perfect balance between function and form.

SAMSUNG

HOW TO

MAKE A WOOD AND CORK COMPOSITE PANEL

Also known as sandwich panels because of the physical interaction of materials, composite panels are widely used in construction and boat building. Wood and cork composite panels are extremely useful for when you want a wood appearance with lighter weight and greater stability than actual wood will give you.

MATERIALS

- Wood veneer (without paper backing)
- 3.2 mm (⅛ in) thick cork core material
- E-glass cloth, around 170 g (6 oz) weight
- Marine-grade epoxy
- Epoxy filler or thickener (if your veneer is porous)

MACHINES

- Disc cutter
- Table saw or bandsaw
- Vacuum pump system

TOOLS

- Butcher's paper or newspaper
- Roller cutter or sharp scissors
- Disposable calibrated measuring cups (for mixing epoxy)
- Mixing stick – plastic, glass or ceramic
- Squeegee – Kurt uses a one-piece unit 15.3 cm (6 in) long
- Clear plastic sheet for release
- Vacuum bag
- 2 flat panels of melamine or pdf 0.6–1.3 cm (¼–½ in) thick, the same size as the veneer you're making. These should have rounded edges to prevent damage to the vacuum bag.
- Clamps
- Sanding materials and tools
- Cork board (or a chopping board you don't mind marking)
- Leather blade or sharp craft knife
- Multi punch
- Mallet
- Cloth
- Stitching pony or vice (optional)
- Leather needles × 2

SAFETY GEAR

- Thin acetone-resistant gloves

METHOD

1. Begin by preparing your work surfaces. You'll need one large work surface (or two smaller ones) for the panel assembly and epoxy mixing, and one fixed location for the vacuum bag.
2. Cover the work area where you're going to do your gluing in layers of butcher's paper – the layers can be discarded as you apply glue, to minimise mess and excess.
3. Use the roller cutter or scissors to trim the veneer and cork to 2.5 cm (1 in) greater than the finished size of your panel. Cut two pieces of veneer and two of cork in this fashion.
4. Use the disc cutter to cut the e-glass cloth to 1.3 cm (½ in) greater than the finished size of your panel. Cut two pieces of e-glass cloth in this fashion.
5. Lay one piece of cork on the butcher's paper, then place one layer of e-glass cloth on top. Straighten the weave by pulling in alternate directions against any waves until the material sits squarely across the cork.
6. Put on the protective gloves, take the disposable cups and mixing stick and carefully mix your epoxy according to the instructions. The exact amount you'll need to mix will vary based on both the porosity of the cork and the openness of the fabric weave.
7. Now pour one-third of the epoxy onto the fabric in a spiral form. Working steadily, use the squeegee to spread the epoxy, moving from the centre to the edges. You'll need to hold the fabric down with your spare hand as you move the squeegee, to avoid moving the glass cloth around on the cork. Aim for full coverage with a minimum of epoxy. Use the remaining epoxy to fill in any gaps, until the entire surface is covered.

HOW TO

MAKE A WOOD AND CORK COMPOSITE PANEL

8. Place one of the prepared veneer pieces on top and press it into place, then take a piece of butcher's paper and place that on top.
9. Grab the upper and lower sheets of paper, making sure you are securely holding the work within, and flip them over.
10. Discard the top paper sheet, then place the second piece of e-glass cloth over the exposed cork, repeat the epoxy application process and put a second sheet of veneer in place.
11. Now put a clear plastic sheet over the top, flip the assembly over, remove the paper and place a second clear plastic sheet on top.
12. Place the whole assembly into the vacuum bag, with the melamine panels above and below. Be sure the rounded edges of the panels face out away from the assembly.
13. Seal the bag by rolling up the open edge and clamping it.
14. Apply the vacuum pump, check that everything is in place, then check the epoxy instructions to determine how long to leave the assembly in the bag.
15. When the glue-up is complete, remove the panel, use a table saw or band saw to trim it to its finished size, and sand it in preparation for finishing.

Step 13

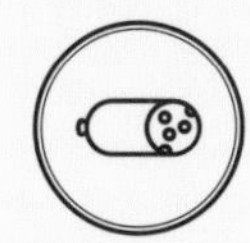

GUBBEEN SMOKEHOUSE CHORIZO

Butchery

Chorizo is tasty stuff. A fermented Spanish sausage made from coarsely chopped pork and pork fat then seasoned with pimenton, smoked paprika and salt, there are hundreds of regional variations on the recipe throughout Spain, Portugal and Mexico. Some call for sweet paprika, some hot, while others include garlic and herbs. Fresh chorizo can be fried or grilled and eaten on its own, added to stews and pies, or simmered in liquids like tomato sauce or cider, and will give heart to any dish. When cured, it will last for months and can be sliced and eaten directly from the sausage. This chorizo is made by Irish charcutier and farming Renaissance man Fingal Ferguson, who raises pigs and butchers them using knives he makes himself. He also smokes the chorizo in a smokehouse he built himself, on a farm that has been in his family for five generations.

MEASUREMENTS	120 g – 1 kg (4 oz – 2 lb 3 oz)
MATERIALS	Black pepper, casings, chilli pepper flakes, curing salt, dextrose, garlic, golden syrup, hot paprika, lactic culture, lemon, oregano, salt, sweet pimenton, pork
KEY TOOLS	Knives
KEY MACHINES	Mincer, sausage machine, smoker
TIME TO MAKE	1 week (fresh); 2–5 weeks (cured)
LIFESPAN	3 months +

FINGAL FERGUSON — Charcutier [West Cork, Ireland]

Fingal Ferguson first tasted chorizo in Andalucia, Spain, as a child. His maternal grandfather lived in Jimena de la Frontera – all white walls and terracotta tiles – and he visited regularly. Food and smells and moments have melded together in his memory, but Fingal's first taste of chorizo was most likely in a tapas bar, as Spanish enthusiasm for life and food spilled out onto the streets.

'My grandfather had strong connections with the locals in Andalucia. He participated in the local tradition of the *matanza* (a ritual pig slaughter) many times, and he always had sharp knives in his house. I loved to handle them, and help my grandmother in the kitchen. These early experiences with animals and meat, and an exposure to old recipes and tradition, made a lasting impression on me and have stood by me as I learned my trade,' says Fingal.

Fingal grew up at Gubbeen, a 250-acre (101 ha) working farm in West Cork that rears pigs, cows, poultry and geese. The farm produces meat and eggs for its own use, as well as for supply to the local markets and to shops, delis and restaurants around Ireland.

'Gubbeen is very close to Mizen Head, which is the most southerly point in Ireland, and in fact we're the most south-westerly cheese dairy in Ireland. This means we're blessed with early grass and clean, salty air.'

Fingal's father Tom runs the farm, while his mother Giana makes the cheese. His sister Clovisse grows a small biodynamic herb garden.

And Fingal? He runs the smokehouse.

Its story goes back to 1989, when a friend in the neighbouring village of Goleen, Chris Jepson, built his own fish smoker. Chris helped Gubbeen Cheese become one of the first dairies smoking cheese in Ireland when he started smoking their cheese. Many years later, Chris helped Fingal and Tom build a smoker similar to his own at Gubbeen.

'When I was young one of my jobs was going back and forth to Chris. He always had bacon hanging from the roof of his smoker, and he showed me how to salt and smoke it.

'Our smokehouse is made out of wood, blocks and stone; it's very simple but very solid in design. It's adjacent to the dairy and has two rooms, one where we put the cheese and hang the cured meat for smoking, and the other where we light the kiln.'

Fingal started off curing bacon for the family. Soon friends were asking for it, and slowly but surely it became a commercial endeavour. Fingal started reading about charcuterie and making connections back to Spain and researching chorizo recipes.

Fingal makes two types of chorizo – a fresh, soft one for cooking and a fully cured one for eating as is.

'We use between 16 and 18 pigs a week here at the farm. We butcher on Monday, make the spice ingredients up and mix them through the meat on Wednesday, then on Friday we mince the mixture again to give it a coarse texture, then stuff it into the casings using our sausage machine.

'Then they go into the smoker, although it's not smoking when they go in; we begin by bringing the temperature to 28 degrees to trigger a 24-hour fermentation process. We then smoke the chorizo for another 24 hours. The cooking chorizo is nearly ready at that stage; the cured chorizo goes into the salami ageing room where it hangs until it's lost one-third of its weight. This takes around two weeks for our skinnier salamis; the larger ones can take five weeks.'

Many things at Gubbeen are made seasonally but, due to demand, chorizo is produced all year round, by hand.

'Hands are important, having this connection with the ingredients – the texture and tackiness and stickiness and how cold or warm the meat is. If you want to become very good at what you're doing and really understand the meat and the process, you have to feel it.'

“

Hands are important, having this connection with the ingredients ... If you want to become very good at what you're doing and really understand the meat and the process, you have to feel it.

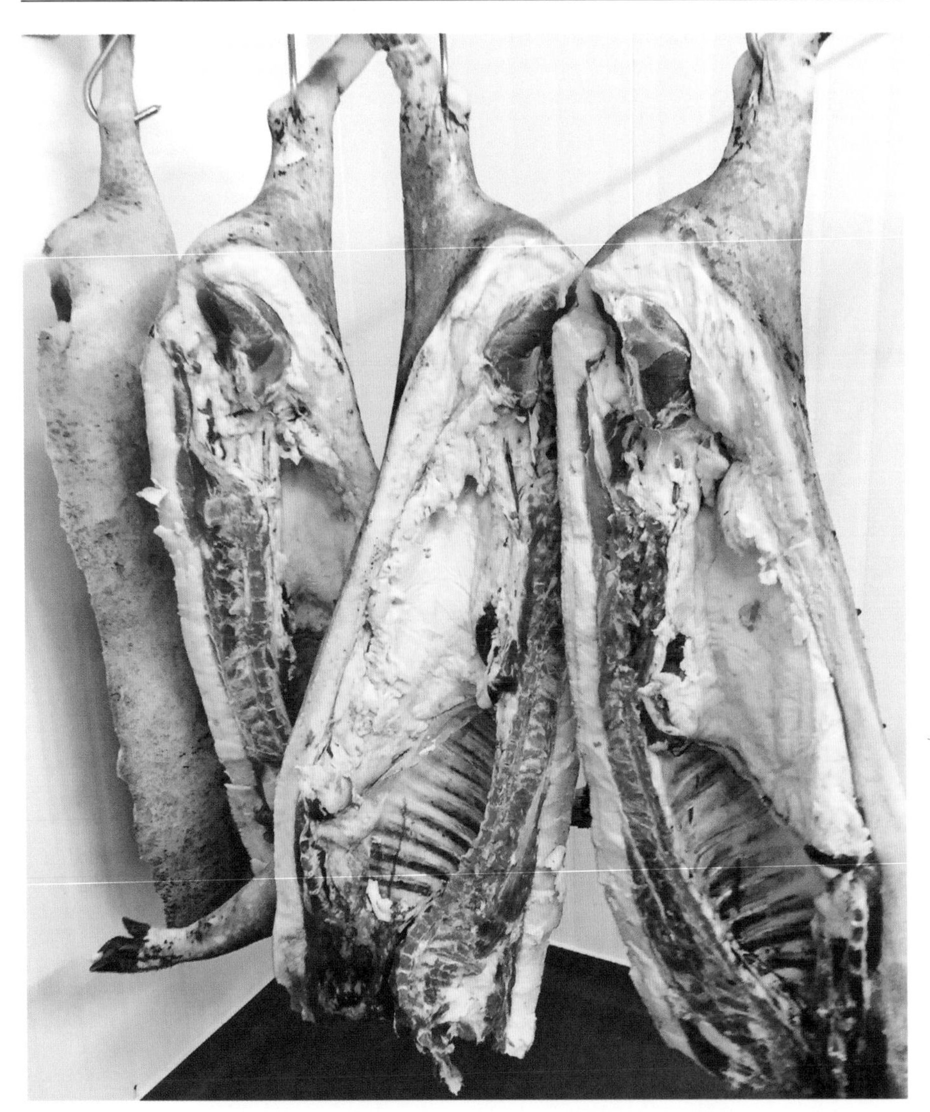

CARING FOR YOUR CHORIZO

Fresh chorizo for cooking must be kept in the fridge and used within a week. Cured chorizo should be stored in a cool dry place and will last for up to three months. Once opened, cured chorizo should be kept in the fridge, and eaten within the fortnight.

HOW TO

MAKE CHORIZO

This home recipe is adapted from the book Gubbeen – The Story of a Working Farm and its Foods *by Giana Ferguson. The recipe makes 15 18-cm (7.1 in) long chorizo of ½ cm (0.2 in) diameter.*

INGREDIENTS

- 5 kg (11 lb) pork, ideally with a 3:1 meat-to-fat ratio – ask your butcher to mince it very coarsely or, if you can mince, use an 8 mm (⅓ in) plate or bigger
- 1 pinch lactic culture
- 95 g (3 oz) salt
- 2 teaspoons curing salt (optional; curing salts contain nitrates, which are essential for commercial production but can be done without at home)
- 2 tablespoons dextrose
- 4 tablespoons sweet paprika
- 4 tablespoons hot paprika
- 1 teaspoon ground black pepper
- ½ teaspoon dried oregano
- 2 tablespoons fresh garlic, minced or finely chopped
- ½ teaspoon chilli flakes
- 5 tablespoons golden syrup (optional; Fingal uses it to marry the smoked flavour with the heat of the spices)

For the casings

- 5 hanks 45–50 mm (1½–2 in) diameter salted natural beef casings
- 1 lemon, quartered and crushed
- 5 teaspoons bicarbonate of soda (baking soda)

TOOLS

- Mincer
- Plastic container
- Bucket of water
- Sausage filler with stuffing nozzle
- Butcher's twine
- Scissors
- Pin

METHOD

Preparing the meat

1. Combine all the flavour ingredients with your minced meat.
2. Vigorously mix everything together with your hands until the meat starts to feel tacky and sticky. Keep the meat below 6°C (43°F).
3. Take about one-third of the meat mixture and mince it again through a 3.5-mm (1⁄10 in) plate. This finer mince acts as a binder for the coarser meat; go ahead and mix it back through the coarser meat.
4. Marinate the mixture by patting it down firmly into a plastic container, then cover it and chill it overnight.

Preparing and filling the casings

1. Wash all salt off the casings and leave them in a bucket of cold water overnight with a piece of crushed lemon to keep them fresh.
2. The next day, rinse the casings by running a cup of cold water into the open end, then gently working it through to the other end with your fingers. This will loosen them and remove any tangles and knots.
3. Leave the casings in a bowl filled with tepid water with the open end hanging over the side.
4. Add the bicarbonate of soda (baking soda) to the water. This will make the casings soft and slippery, which will help when running them onto your stuffing nozzle.
5. Insert the meat mixture into the sausage filler a bit at a time, packing it down to prevent any air bubbles in the mix.
6. Attach the first casing to the stuffing nozzle, then fill the casings by inserting the nozzle into the casing and squeezing the mixture in.

HOW TO

MAKE CHORIZO

7. Once the casings are filled to your desired length, secure them with butcher's twine and tie off the ends with a double knot.
8. Weigh each sausage and make a note of the fresh weight on a label for each one.
9. To remove any air bubbles, stab the casings with a pin multiple times, then hang the chorizo on hooks or over sticks to dry. (Fingal makes hooks from fencing wire, which can be bent by hand. More urban types can utilise wire coathangers.)

Incubation, smoking and drying

1. Hang your casings somewhere warm to drip dry. Ideally you want the chorizo to warm up slowly to 28°C (82°F) in a humid place to accelerate fermentation, and so that the casings don't harden. After 24 hours they are ready to be smoked.
2. How long they take to smoke depends on your smoker – in the Gubbeen smoker, 24 hours is perfect.
3. Next, hang them in an airy place like a pantry where conditions are as near as possible to 75 per cent humidity and 15°C (59°F). If mould appears on the surface of your chorizo, sterilise them by wiping the mould off with a diluted cooking alcohol like brandy or vodka.
4. When the chorizo have lost a third of their weight, which will take 2–5 weeks, they are ready to try.

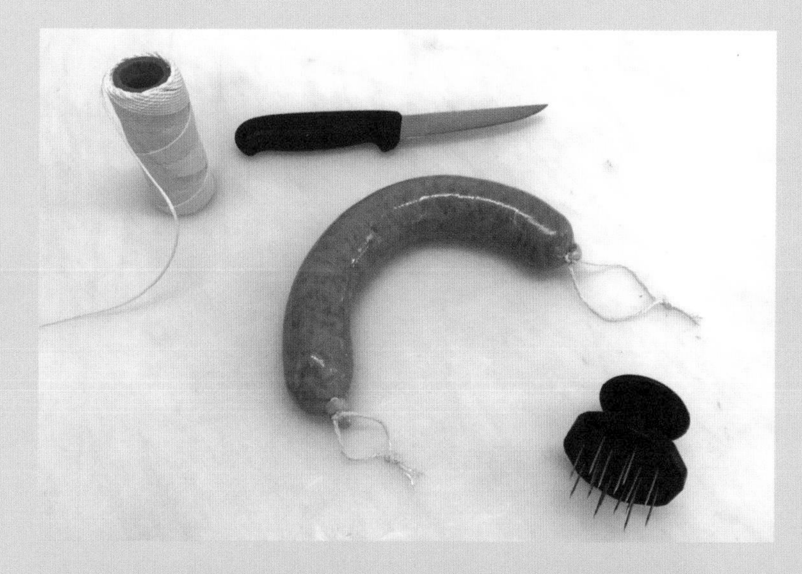

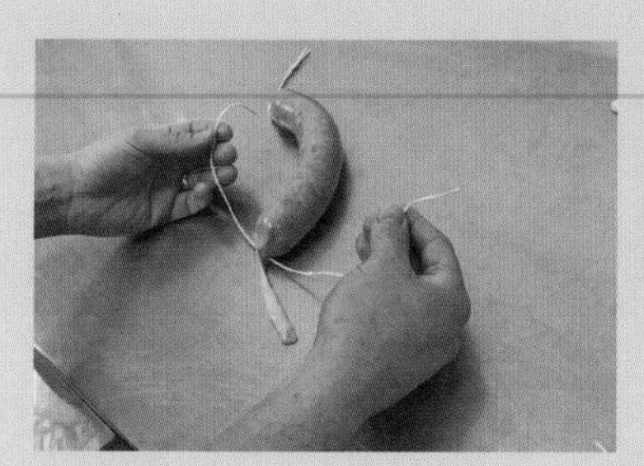

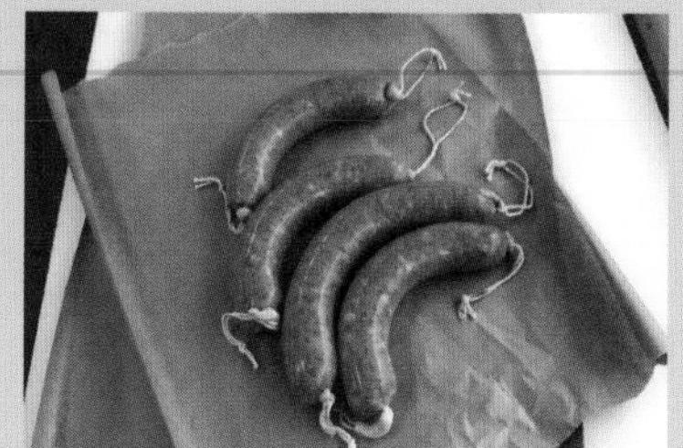

Step 7

SMALL BATCH
PREMIUM GIN
LIQUOR CO.
PROHIBITION
69% ALC VOL
500ML
EST. 2015
GIN
BATHTUB
CUT GIN
ORIGINAL
PRESCRIPTION FORM FOR MEDICAL LIQUOR
In Prohibition times the bathtub was the clandestine vessel for bootleg gin manufacture, soaking juniper and other botanicals to create 'bathtub gin'. Prohibition Bathtub Cut pays homage to this heritage, with a secondary maceration of juniper post-distillation, cut from the still, undiluted at 69%. The result is a punchy, yet sophisticated overproof gin... with the true spirit of prohibition.
PRODUCT OF AUSTRALIA 69% ALC./VOL.
CONTAINS APPROX. 28 STANDARD DRINKS
PRODUCED BY PROHIBITION LIQUOR CO.
PO BOX 232, STIRLING
SOUTH AUSTRALIA 5152
DRINK BATHTUB
CUT RESPONSIBLY.
TREAT WITH CARE,
IT MAY BITE.
9 369999 063184

LIQUOR CO.
PROHIBITION
EST. 2015

BATHTUB CUT GIN

Distilling

In the United States of America, from 1920 to 1933, the production, importation, transportation and sale of alcoholic beverages was banned under Prohibition. Some people toed the dry crusaders' line, but many didn't, and bootlegging operations quickly sprung up across the land. Gin was a popular choice, as it could be made quickly and cheaply – often in bathtubs. This brew is a modern nod to gin's bootlegging past. A 69 per cent proof gin infused with 13 flavoursome botanicals, it is made in a former cold store in the Adelaide Hills that dates back to the early 1900s. It is brewed by the clever chaps at the Prohibition Liquor Co., who have been creating small batches of award-winning premium gin so rich and strong it'd have made the hair on your great-grandfather's neck curl with illicit pleasure.

MEASUREMENTS	Batch: 150 litres (568 gal); Bottle: 500 ml (1 pt)
MATERIALS	Almonds, base alcohol, cassia bark, coriander seed, ginger root, grapefruit, green tea, juniper, lavender, orange rind, orris root, star anise, vanilla, wormwood
KEY TOOLS	Stainless steel drums
KEY MACHINES	Peristaltic pump, stainless steel still
TIME TO MAKE	4 days
LIFESPAN	50 years +

ADAM CARPENTER AND WES HEDDLES — Bootleggers [Adelaide Hills, Australia]

It was a series of conversations over gin (what else?) that got mates Adam Carpenter and Wes Heddles excited about the prospect of doing their own distilling. They'd watched with interest as the Adelaide Hills became a hotbed for small craft beer breweries and spirit distilleries, but noted that none were making the world's fastest growing spirit: gin.

This, in combination with the realisation that each loved gin as much as the other, made the next step a logical leap: in early 2015, the Prohibition Liquor Co. was born.

'We distil in the old Gumeracha cold stores in the Adelaide Hills in South Australia, a region which was common for apple and cherry orchards. The cold stores have a rich history of storing fresh local produce for distribution around the country,' says Adam.

Now Gumeracha is the site of Applewood Distillery. It's where their master distiller, Brendan Carter, is based, and it's where Wes and Adam did much of the research for their Bathtub Cut Gin. They spent weeks tasting gins from all over the world to work out what they wanted from their own. Then they began experimenting with the botanicals.

Other gins in their line took weeks of testing, tasting and refinement to perfect, but the Bathtub Cut Gin took just one day. They worked with a 1-litre (1 qt) micro still, pounding and cutting and mushing botanicals and distilling them in alcohol until it was just right.

Adam and Wes's set-up would be the envy of any Prohibition-era bootlegger. The bulk base spirit is bought in from a local company that turns grape lees (residual deposits that precipitate to the bottom of a vat of wine after fermentation and aging) from winemakers in the nearby Barossa Valley into high-grade, usable product – in this case, ethanol. The botanicals are bought in, too. But the maceration and distilling are all done by hand, on site.

Maceration involves soaking the natural botanicals (primarily juniper, which is what makes it 'gin') in the spirit to impart the flavour and create the signature notes of the gin.

'We do this in a particular order, from those that take the longest to macerate down to the shortest. Juniper, coriander seed, ginger and orris root are the basis of a London Dry Gin; the wormwood, star anise and cassia bark add the spicy, savoury flavour components,' says Wes.

'Then the citrus elements add acidity, and the almonds and vanilla add an oiliness to make it smooth, sweet and silky. Finally, the green tea works to settle the flavour down, and lavender brings a soft, floral finish.

'After a few days of maceration, we pour the spirit and botanicals into the 300-litre (80 gal) pot still and then we add a few final botanicals, which only activate in the still. Then it's time to start the electric still. Once the alcohol reaches 78°C (172°F), it starts flowing through the columns,' explains Wes.

Wes, Adam and Brendan wait for the tops – the oils that float to the top – to pass through the still first, putting aside the first few litres of liquid to use for cleaning spirit or similar.

'We let the still run for several hours, tasting it regularly until the run is nearing the tails. We can tell this is the case when the gin starts to taste bitter. We cut the run and discard the tails portion of the run.'

Next, the Bathtub Cut brew is gently diluted down to 69 per cent alcohol volume, or 138 per cent proof, with distilled water, then decanted into 50-litre (13 gal) stainless steel drums. Once it's in the drums, it's given a secondary maceration.

'We add juniper berries back into the drums for a quick bath of secondary maceration. This brings a beautiful golden colour and additional juniper hit into the spirit.'

It's then shipped in bulk spirit to Prohibition Liquor's bottling operation in Adelaide, where it's decanted via a peristaltic pump into bottles, labelled, sealed and packed into boxes for shipping to distributors around the world.

CARING FOR YOUR BATHTUB CUT GIN

Store your Bathtub Cut Gin in a cool, dry environment. Keep it sealed – unless you want to experiment with the impact of oxidisation (exposure to air) on the gin's flavour profile. If this is the case, put a pourer in the bottle in between drinks. It will last just as long, but you may notice the botanical notes mellow over time.

> We distil in the old Gumeracha cold stores in the Adelaide Hills in South Australia ... The cold stores have a rich history of storing fresh local produce for distribution around the country.

HOW TO

MAKE BATHTUB GIN

Sure, professional gin makers extract the flavours from their botanicals through distillation, but it's possible to make gin simply by steeping your botanicals in a base spirit. You'll definitely need vodka and juniper berries, but beyond that, experiment with your favourite flavours from your spice cabinet, herb garden or edibles in your ecosystem.

INGREDIENTS

- 750 ml (25½ fl oz/3 cups) quality vodka
- 2 tablespoons juniper berries
- Your chosen botanicals in relatively small quantities. Here we've used:
- 1 teaspoon coriander seeds
- 1 star anise pod
- Half a cinnamon stick
- 1 vanilla bean pod
- 2 peppercorns
- Orange peel (pith removed)
- Lemon peel (pith removed)

TOOLS

- Mason jars (or similar)
- Boiling water
- Sieve
- Muslin cloth or a paper coffee filter
- Bottles, for bottling

METHOD

1. Sterilise the clean mason jars with boiling water.
2. Put all the botanicals except the orange and lemon peel in a mortar and pestle and gently grind them, then put them in the jar.
3. Fill the jar up with vodka, then put the lid on and leave it to infuse in a cool dark place for 24 hours. Give it a gentle shake every now and then.
4. Open it up and taste the infusion (heck, why not?), then add the fresh peel, along with a little more of any botanicals whose flavour you want to boost.
5. Put the lid back on and leave it for up to 24 hours – just beware of leaving it too long and over-infusing the mixture.
6. Taste the gin again. If you're happy with the taste, strain it into another sterilised jar of the same size, catching the botanicals in the sieve.
7. If there's still some sediment left, strain the gin again, this time through some muslin or a coffee filter.
8. Put the lid on the jar and leave your gin to sit for a couple of days, then filter out any remaining sediment.
9. Bottle your gin – you might even want to create your own label.
10. Drink a toast to the Prohibition era.

Step 2

Step 3

Step 6

Step 8

SPLIT CANE FLY FISHING ROD

Rod making

Incredibly strong yet almost infinitely flexible and remarkably waterproof when heat-treated, split cane was the material for fly-fishing rod making for hundreds of years. After World War II, factory production of fibreglass, and then carbon fibre, rods nearly blew split cane rods out of the water. But committed fly fishers prefer the look and feel of split cane rods. Fishing rod maker Edward Barder has been at his craft for more than two decades, making light trout rods, salmon rods and coarse rods. This one is a split cane fly fishing rod – perfect for catching trout. When he's not in his workshop patiently splitting cane and crafting made-to-order rods from seasoned bamboo, Edward can be found on the banks of the Kennet River, a famous chalk stream at the end of the garden just outside his workshop.

MEASUREMENTS	2–2.6 m (6½–8½ ft) long
MATERIALS	Tonkin cane from remote China (rod), flor-grade cork from Portugal (handles), English-made Pearsall's silk (whippings that secure ferrule splints and guides), highly figured seasoned olive wood (reel seats), 18 per cent nickel silver (handle fittings and ferrules), guides/rod rings, re-bronzing agent, yacht varnish
KEY TOOLS	Blowtorch, files, mallet, pen knife, ruler, sandpaper
KEY MACHINES	Bamboo drying oven, bamboo mill, glue binder, lathe, precision metal lathe
TIME TO MAKE	100 hours +
LIFESPAN	Heirloom quality

EDWARD BARDER — Split cane rod maker [Berkshire, England]

When your father is a well-known angler who wrote books about how to fish, and you spent your childhood experimenting with his rods in the lakes and streams of Berkshire, it's no surprise that you go on to become a veritable split cane rod perfectionist, and your creations some of the most sought-after fishing rods in the world.

The waiting time for an Edward Barder rod used to be about 16 weeks; now it's around two and a half years.

'Our output capabilities determine the turnaround time – we expect to finish about 25 rods each year, maybe 30 at a push. We make virtually all parts of the rod here, so that we are completely in charge of supply, quality and design.

'We used to buy in certain parts, such as ferrules, but we took the decision years ago to equip ourselves to be independent. Now, the only things we buy in are rod guides [rings] – but even then we customise them ourselves – the rubber butt caps for our coarse rods, and our leather rod bags, which are made by a local craftswoman.'

Edward starts by making the rod blank, which forms the core of the fishing pole. He takes a Tonkin bamboo pole, which is approximately 5–6 cm (2–2½ in) in diameter and classed as a 'grade-A' pole for its colour, size, straightness, heft and cleanliness, and splits it into between 12 and 18 strips using a farmer's pen knife and mallet. He stands on a chair to reach the top of the 3.7 m (12 ft) pole, gravity and force taking hold with each firm and steady rap of the mallet.

The strips have knots at intervals, which need to be straightened out. Edward files as much as he can, then heats the bamboo over a low-pressure gas jet; a soft cloth protecting the hand that is holding the pole as it heats.

'We use the heat to temporarily make the bamboo soft so it can be straightened as it cools,' explains Edward.

'Heating the bamboo also drives a lot of moisture out of it. This is important, as, if this moisture isn't removed the bamboo will be

"

Nothing beats seeing the pleasure on the face of the client when you hand them a rod that you know they've saved up for and will cherish all their life.

soggy and won't recover properly from being bent. We carefully work the bamboo over a low-pressure open flame – this measured and tempered heat treatment shrinks the bamboo, and helps it to become denser and more stable.'

Next, Edward tapers the bamboo strips into triangular strips using a milling machine. Then he glues them together to create the rod blank's hexagonal form, with the strongest fibres on the outside.

'The milled strips are soaked in a special heat and waterproof glue and then wrapped twice, spirally, in a binding machine to squeeze out any excess glue. This ensures a strong, tight fit and a true, straight rod.'

Next, the outer shell of the bamboo and any glue residue is very carefully sanded off. Edward does this by hand, turning the blank as he works sandpaper over it. He fits the nickel silver ferrules, uses a lathe to turn the cork handle, fits the reel seat, and attaches the guides, which are made of pure silk.

Lastly, Edward varnishes the rod using a high-grade yacht varnish to create an immaculate, water-resistant finish.

For a craft with a purpose as delicate and as synchronous with the workings of the natural world as Edward's, patience and calm are hugely valuable personal resources.

'We're using natural materials, and sometimes things such as bamboo straightening or varnishing can't be hurried. Nothing beats seeing the pleasure on the face of the client when you hand them a rod that you know they've saved up for and will cherish all their life.'

CARING FOR YOUR SPLIT CANE FISHING ROD

Never yank at a fixed object or fish, or over-strain a rod in casting. Using a split cane rod outside the limits of its specification may result in the bamboo fibres becoming stretched beyond their capacity to return to their original position. A moment's carelessness can ruin a rod that would give a lifetime of good service if handled correctly.

Make sure there is plenty of room around you when assembling and taking down a rod. Never twist the rod when assembling it or taking it apart. Always hold the rod sensibly and securely, pushing and pulling in a straight line when inserting and separating ferrules. A helpful mantra is 'Rod together, hands together – rod apart, hands apart.'

Clean your rod often. Wipe it down with a damp cloth, then a dry one. Wash the cork handle from time to time, using warm water and a drop of washing-up liquid. Keep the ferrule slides clean by carefully wiping with a clean cloth. Start with a damp corner and finish with a dry one. Lightly rub the male slide with hard hand soap if it needs lubricating, but only when absolutely necessary. Store the rod in its bag with the thin ends facing upwards, and hang it from a hook in a dark dry place. Keep fly lines clean by wiping them down with a damp cloth and soapy water. If they are run out and stretched before each trip, they'll perform more effectively.

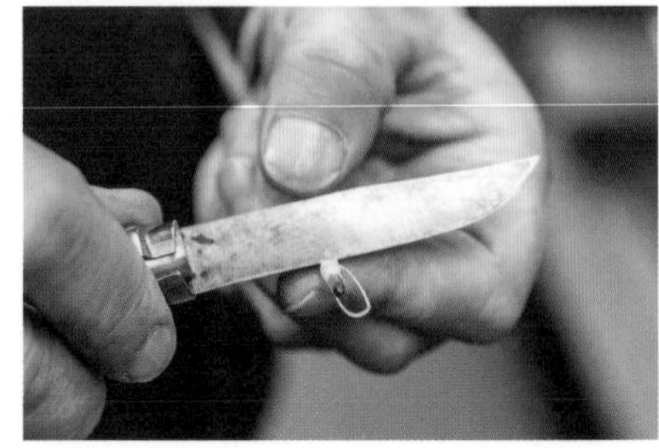

HOW TO

LAND A FISH

Edward has been fishing since he was a boy, and he has landed more fish than ... Well, chances are he's landed more fish than most. Catching a fish on the end of a line is one thing; landing it is another altogether. Here are his tips for how to successfully land a fish.

1. When playing and landing fish, the rod butt should be at or close to 90 degrees to the line beyond the tip ring. As more pressure is applied, the lower, stronger part of the rod will take the strain, relieving the relatively fragile tip section.
2. During protracted struggles with big fish, try holding the rod with the rings facing upwards. This is easier than it sounds, and distributes heavy loads on both sides of the rod.
3. Use a landing net with a sensibly long handle. A short-handled net will force you to hold the rod almost vertically while landing a fish. This places unnecessary strain on the rod top, and should be avoided.

HONEY
FLORA

WARRE WRAPS

Beekeeping

Beeswax is incredible. It's abundant, doesn't dissolve or change its properties in water, and its chemical makeup is so stable that it lasts thousands of years. Humankind has harnessed its properties in many ways throughout history: in skin balms, beauty treatments and medicines; for candle, soap and crayon making; for embalming and waterproofing; and for food preservation, as these beeswax-infused food wraps attest. Made by dipping pieces of vintage or recycled cotton fabric in a fortifying mixture of beeswax, pine resin and jojoba oil, they can be used to cover and wrap food – think sandwiches, cheese, bread and the like – time and time again. They're made by two homesteading mamas in the green outskirts of Sydney, Australia, who keep their own bees and harvest their own wax.

MEASUREMENTS	Extra small: 14 × 14 cm (5½ × 5½ in); Small: 20 × 20 cm (8 × 8 in); Medium: 26 × 26 cm (10 × 10 in); Large: 33 × 33 cm (13 × 13 in); Extra large: 43 × 43 cm (17 × 17 in)
MATERIALS	Beeswax, pine resin, jojoba oil, 100% cotton fabric
KEY TOOLS	Chisel, clothes pegs, cooking pot, iron, lint brush, pinking shears, mortar and pestle, string
KEY MACHINES	Stovetop
TIME TO MAKE	4 hours (batch of 140 wraps)
LIFESPAN	1 year +

EMILY GIMELLARO AND TANIA DICKSON — Backyard beekeepers [Wilton, Australia]

Friends Emily Gimellaro and Tania Dickson know a thing or two about bees. The buzzy creatures pollinate a third of what we eat, communicate via dance and, as these women will enthusiastically tell you, produce many amazing things. Emily and Tania make amazing things, too: lip balms, salves and moisturisers, sunscreen, food wraps, furniture polish, spoon butter, candles.

Of all the things they've managed to produce from beeswax, it's the reusable food wraps that they have decided to pursue commercially.

'We try and minimise single-use plastic as much as possible at home, and it's nice to make something that encourages others to do the same,' says Tania.

The wrap making process is relatively simple. They start by cutting the fabric to size with pinking shears, which stops the fabric from running. They give each piece a good shake, and remove any excess fluff with a lint brush. Then the fabric is ironed and piled neatly, ready for waxing.

To prepare the wax mixture, the pair chisel the beeswax into manageable chunks and grind the pine resin in a mortar and pestle. These, along with the jojoba oil, are melted in a big pot on the stovetop. The mixture is then carefully poured into a purpose-built receptacle.

'We hold the wraps in our fingers to dip them in the wax mixture. We've found we're more accurate with our fingers than with pegs! Once they're dipped we peg them on the lines we have strung up across the shelves and cupboards in our workroom. The wax dries incredibly quickly; within a few minutes, they're ready for folding and packaging,' explains Emily.

Emily and Tania spent about four months refining their production processes to make them as streamlined and sustainable as possible. As well as enabling them to simply dip a swathe of fabric in the mixture (rather than the more intensive pouring and brushing used in the smaller-scale oven production method), the custom dipping receptacle minimises stovetop time, and therefore energy use. It also eliminates the reams and reams of baking paper demanded by the oven method, which irked them.

Wrap-making also allows Tania and Emily to indulge in one of their favourite pastimes: second-hand fabric shopping.

'We visit opportunity shops and vintage stores, buy bulk seconds or offcuts online, and even get some from our mothers, who are both quilters and often have spares. We have a huge stash of fabrics in our work cupboards, and we love sifting through them to choose patterns for the batch we're about to start working on,' says Tania.

'We love the variety in our fabrics. The only must is that the fabric must be 100 per cent cotton. We're experimenting with botanical dyes at the moment,' adds Emily.

They named their wraps Warre Wraps, as an homage to the type of hives they keep their bees in. A warré (pronounced *war-ray*) hive is a vertical top bar hive that is made from identical boxes fitted with top bars and stacked one on top of the other. It's designed to mimic one of a bee colony's favourite hive-making spots: a hollow tree. Warré hives are cheap and easy to build, closer to nature than other hive types, and great for smaller spaces – like Emily's backyard, where she keeps them in the chicken run.

Emily's hive is home to 60,000 or so bees, who produce 3–4 kg (6–8 lb) of wax a year. And then there's the annual honey harvest – all 40 kg (88 lb) of it, which the family eat, share with friends and trade.

'I chose to bee keep in a warré hive because I feel that it is the closest way to nature for the bees. I love how gentle the bees are because they are busy building their own comb. There are no plastic frames, no queen excluders, no feeding sugar syrup. It's just bees in a box doing their thing,' says Emily.

CARING FOR YOUR REUSABLE FOOD WRAPS

Use your reusable food wraps in place of plastic wrap for anything except raw meat, greasy food or acidic food. Between uses, rinse the wraps in cool soapy dishwater, wipe them clean with a cool damp cloth and let them air dry. Store them in a cool place out of direct sunlight – either on a clean shelf in your pantry or in a container. It's okay to gently fold them to fit in a container if need be.

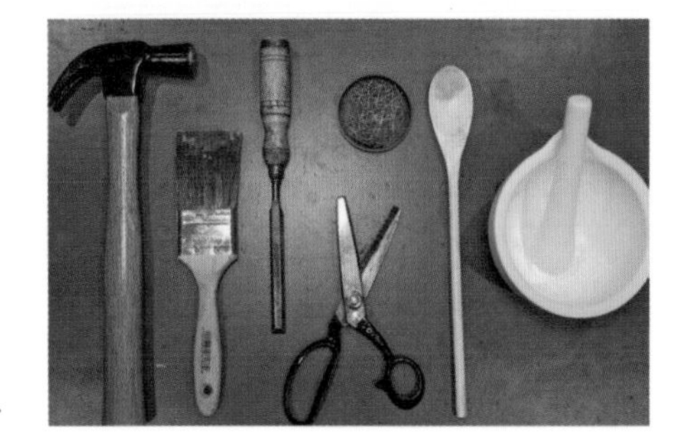

> We try and minimise single-use plastic as much as possible at home, and it's nice to make something that encourages others to do the same.

HOW TO

MAKE REUSABLE FOOD WRAPS

INGREDIENTS

- 100% cotton fabric – size and pattern of your choice
- Water (for double boiling)
- 1 cup beeswax (chunked or pastilles)
- ¼ cup tree resin (pine is good)
- 2 tablespoons jojoba oil

TOOLS

- Pinking shears or sharp scissors
- Oven
- Medium-sized cooking pot
- Mason jar
- Oven gloves
- Clean paintbrush (or pastry brush)
- Newspaper or an old towel
- Pegs
- Coathanger or indoor clothesline

METHOD

1. Cut your fabric to size – the sizing of the Warre Wraps is a great guide. You can use any scissors for this, but pinking shears are best as their zigzag cut prevents the fabric from fraying.
2. Heat your oven to 150°C (300°F).
3. Half fill the cooking pot with water, then set it on the stove to boil.
4. Place the beeswax, resin and jojoba oil in the mason jar, put the lid on tightly, then immerse the jar in the water. When the water is boiling the mixture in the jar will start to melt.
5. While this is happening, line a baking tray with the waxed paper and lay a single piece of fabric on the tray.
6. When the mixture has fully melted, put on oven gloves, remove the hot jar from the boiling water, and carefully drizzle the beeswax mixture over the fabric. It will harden quickly, but you'll have a chance to smooth out the wax at the next stage, so don't worry too much if it's a bit uneven at this stage.
7. Place the tray in the oven for a couple of minutes to reheat the beeswax. Pull the tray out and use the paintbrush to spread the mixture evenly over the cotton's fibres, then return the tray to the oven.
8. Lay some newspaper over your work area, then remove the tray from the oven. Use two clothes pegs to pick the fabric up at the corners and hold it over the newspaper while any excess wax drips off.
9. Peg the fabric on an old coathanger or clothesline.
10. Let your wrap dry for a few hours before using.

PRO TIP

The compostable baking paper used to line the oven tray can be re-used again and again while you make different wraps, as the wax it catches will melt and be absorbed across many pieces of fabric.

Step 1

Step 2

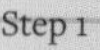

Step 4

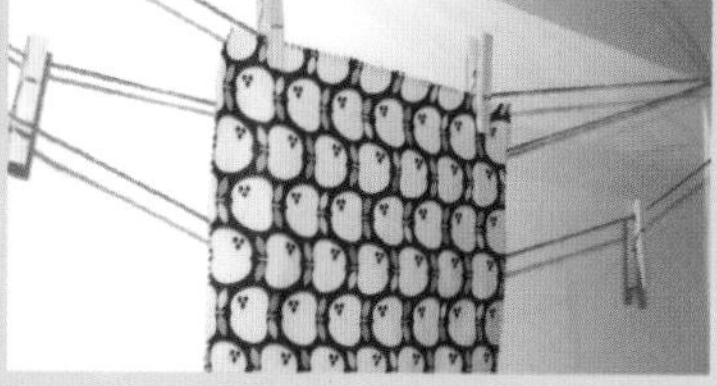

Step 9

Thrones

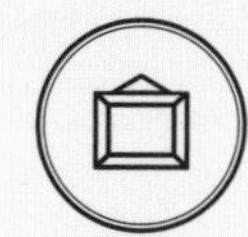

HANDCRAFTED CONSERVATION FRAME

Picture framing

This conservation frame was handcrafted from a recycled fence post in a workshop surrounded by some of the tallest trees in the world, in the Dandenong Ranges, some 35 km (22 mi) east of Melbourne. The frame's maker, Shaun Duncan, has made thousands of conservation frames and framed artworks in them in ways that ensure they'll last hundreds of years. The internal materials used in the frame – the mat board, foam core and glass – are all acid-free. Museum-quality glass protects the work from UV light, while frame-sealing tape prevents any outgassing (the release of gases that are trapped in the timber) from affecting the artwork, which is mounted in a fully reversible way that hasn't altered the artwork a smidge. As any art aficionado knows, this is invaluable when you're looking to maintain and grow an artwork's value.

MEASUREMENTS	48.7 cm × 35.2 cm (19 in × 14 in)
MATERIALS	Acid-free foam core, acid-free mat board, Danish oil, museum glass, recycled timber
KEY TOOLS	Beading tool, chisel, compound mitre saw, dovetail saw, mat cutter, Stanley #5 jack plane, Stanley #78 rebate plane, Stanley #4 smoothing plane, shooting board, table saw, tenon saw
MACHINES	Nil
TIME TO MAKE	10 hours + drying time
LIFESPAN	Heirloom quality

SHAUN C. DUNCAN — Framer [Belgrave, Australia]

Before he started framing, Shaun Duncan worked in an office and couldn't conceive of work as a spiritual discipline. But now, when he's in his workshop, he feels a sense of calm focus that he's never found elsewhere. He says there's not much space for ego when it comes to framing: 'even the finest work must be subordinate to the art'.

Shaun makes the moulding for a frame first, then the frame itself. The moulding is the decorative, angled part of the frame that travels between the glass and the frame's outside edges.

He starts by cutting the timber into lengths using a table saw. He examines each length for quality, and uses a pencil to mark which will be the face, and which will be the outside.

Then Shaun mounts the lengths to a sticking board, which holds the timber in place so he can plane its entire length with one stroke. He squares and dimensions them using a Stanley #5 jack plane, then planes the face of the moulding to a 5-degree bevel, cuts a bead (semicircular piece of moulding) into the outside edge using a beading tool, then cuts a rebate (a recess in the edge of the wood) to create a lip for the frame.

With the moulding complete, it's time to move on to the frame. Shaun cuts the mitres for the frame first. These are the joints, and are made by cutting an angle on the end of each piece of wood and then fastening the two angles together.

'The mitres can be cut by hand, with a compound mitre saw or twin-blade mitre cutter. A clean and precise 45-degree angle is essential for all cuts. This is difficult to achieve by hand, so I neaten them up using a shooting board.'

Now the frame is ready to join. Shaun does this one corner at a time, positioning two lengths in the vice and tracing a line 2 cm (¾ in) from the end of each length. Then, using a dovetail marker, he marks the kerfs (gaps) for the splines (a slat or key to strengthen the corner join), tracing the angle down from the corner to the line he traced earlier. Then he cuts the kerfs using a tenon saw.

'I cut the splines from Snow Gum, going across the grain. They need to be cut to the width of the kerf, which can be difficult. I find using a larger tenon saw for the kerfs, followed by a smaller dovetail saw for the splines, makes it easier.'

'It's good to do a dry run before gluing the corner up. I test to make sure the splines fit snugly into the kerfs. I don't want to have to force them into place, but I don't want them to be too loose, either.'

Shaun uses plenty of glue on the corners. He'll be planing the sides clean later, but even so, he tries not to get any on the face. He glues the four corner pieces together, uses a knife or chisel to trim off the exposed splines, then leaves them to set overnight.

The next day, he glues the two halves together to form the complete frame, then leaves it to dry for another night. The day after that, he planes the sides clean with a Stanley #4 smoothing plane, in preparation for finishing.

'I decided to use Danish oil as the finish for this frame. I applied it using a clean rag and left it to sit for 20 minutes or so before wiping off the excess. Another 24 hours, and it's ready for another coat. If any of the grain has raised in the meantime, the frame can be smoothed back using fine-grade steel wool or sandpaper.'

Once it's completely dry, Shaun gives it a final gentle sand. Then it's time for the framing itself.

CARING FOR YOUR CONSERVATION FRAME

Hang your frame away from direct sunlight. The timber may be touched up with oil or wax and buffed with a soft rag as required.

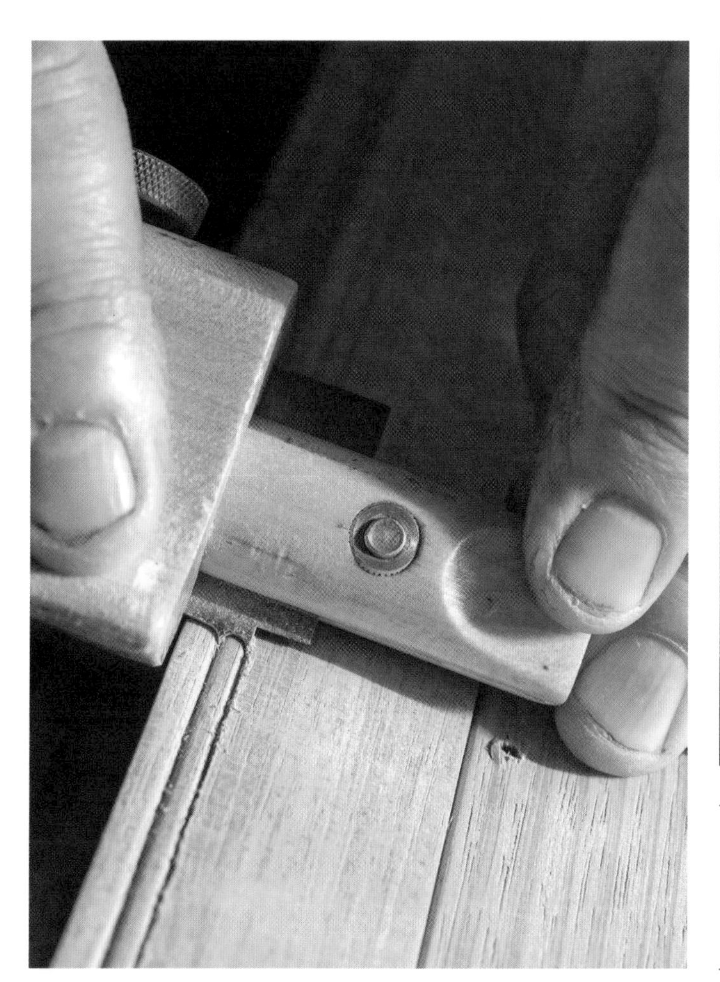

Even the finest work must be subordinate to the art.

HOW TO

FRAME A PICTURE

INGREDIENTS

- Picture frame
- Artwork for framing
- Mat board (acid free)
- Foam core (acid free)
- Museum glass
- Hinging tape (acid free)
- Frame sealing tape (foil backed)
- Brown framer's paper tape
- D-rings with screws
- Hanging wire

TOOLS

- Steel ruler or measuring tape
- Mat cutter
- Stanley knife
- Glass cutter
- Tacks, staples or a point driver
- Drill

METHOD

1. Lay the frame on your work surface, facedown.
2. Measure the artwork and the frame, then subtract the size of the artwork from the size of the frame. The difference between the two is the starting point for working out the size of your mat board. Shaun likes to leave around 6 cm (2½ in) at the bottom and 5 cm (2 in) on the other three sides of a mat board.
3. Mark out the measurements on the back of the mat board with a ruler and pencil.
4. Take the mat cutter and steel ruler and firmly cut through the mat board. If you don't have a mat cutter, a Stanley knife will do (though it won't cut a bevel).
5. Mark up the measurement for the foam core (it will be the same size as the frame), then use the Stanley knife and steel ruler to cut it out.
6. Next, measure the glass. It will be 2–3 mm (1⁄10 in) smaller than the internal dimensions of the frame, to allow room for it to expand.
7. Use a glass cutter to score and snap the glass.
8. Hinge the mat board at the top and mount the artwork to the mat board using the T-hinging technique: attach three strips of hinging tape to the back of the artwork so the ends protrude from the top by about 2 cm (¾ in), then lay three more strips across the protruding ends to fix the artwork to the backing board.
9. To protect the artwork from any outgassing from the timber, seal the interior of the frame with foil-backed frame sealing tape.
10. Now you can insert the glass, matted artwork and foam core (which is used to fill out the back of the frame) into the frame, in that order. Make sure they're facing away from you. Fix the foam core backing in place using tacks, staples or a point driver.
11. Tape up the back of the frame using more of the sealing tape, or brown framer's paper tape.
12. Finally, use the drill to attach the D-rings to the back of the frame, then string the framing wire between them. String the wire as tightly as you can, looping about 5 cm (2 in) through each D-ring and winding the excess back around itself.
13. Your frame is ready to be hung.

Step 4

Step 7

Step 7

Step 8

Step 9

Step 12

Step 12

CAST-IRON FRYING SKILLET

Metal casting

Cast-iron pans are the workhorses of the culinary world. They're virtually indestructible, easily restored if mistreated, and improve with age. This particular skillet is hand cast from 100 per cent recycled iron, and is a browning, searing and shallow-frying wunderkind. But the real showstopper is its long, heat-dissipating handle; you can forget the oven gloves and pick it up with your bare hands. And thanks to a pre-seasoning of organic, kosher-certified flaxseed oil, the pans are ready for action straight out of the box. Prefer to season your own pan? They're made to order by a maker duo who are committed to inserting modern, sustainable practices into a pollutant-heavy industry, and who are keen to ensure that the people who buy their pans get exactly what they want. All you have to do is ask.

MEASUREMENTS	Pan diameter (lip to lip): 26.5 cm (10½ in); handle length: 17.1 cm (7 in); weight: 2.8 kg (6 lb 5 oz)
MATERIALS	100 per cent recycled iron from old steam radiators, auto brake rotors and the like, sand
KEY TOOLS	Hammers, sand mould, sandpaper, skimmers, tongs
KEY MACHINES	Belt grinder, crucible, electric tilt furnace, grinder
TIME TO MAKE	1 week
LIFESPAN	Forever (unless dropped)

JOHN TRUEX AND LIZ SERU — Metal casters [New York, USA]

Not all couples would choose to work together in an enclosed, high-temperature environment, but for metal casters John Truex and Liz Seru, the magnetism is red hot. The pair spend many an evening pouring molten steel into a graphite crucible, making cast-iron skillets.

Borough Furnace has been making pans since 2011. Keen to reinterpret a beloved and iconic cooking tool in a more contemporary form, John developed the skillet's design while studying sculpture at the University of Tennessee.

Borough Furnace's workshop is in a one-time Rust Belt city from the heyday of US manufacturing: Syracuse, upstate New York. The city has modernised and regenerated in recent years. There's an aptness, then, to the fact that John and Liz have developed a manufacturing process that matches 21st century standards of sustainability. It's a progressive thing to do, given that many in their industry still operate in an environmentally detrimental way.

'Traditional foundries consume massive amounts of electricity from a coal power plant to melt their substrate in an induction furnace, or burn coke – a form of processed coal – in a cupola furnace. Source material is typically 'pig iron', iron derived from smelting raw iron ore,' says Liz.

'Pouring is always exciting. Even though we have great safety precautions in place, there's a certain amount of adrenaline that kicks in, being so close to so much heat!' says Liz.

It takes an hour in a furnace and a peak temperature of 1482°C (2700°F) to melt steel from solid to liquid.

'When it's ready, we pour the molten metal into a graphite crucible. Any impurities in the metal float to the top, so we remove them by skimming them off with a skimmer,' says Liz. The impurities are called 'slag', and the process is called 'slagging'.

Then, they wait for the metal to cool to around 1371°C (2500°F), when it can be poured into the waiting sand mould. This is a special, skillet-shaped mould that is formed by ramming sand around a two-part master pattern that is then bookended together to create the casting's shape.

'We typically pour in the evening, then let the casting cool in the mould overnight. The next morning we remove the mould and clean all the sand off of the surface of the metal to prepare it for finishing. The path that the metal takes while travelling through the mould is cut from the casting with a grinder – the "gating". From there, the exterior edges of the skillet are ground smooth on a belt grinder,' continues John.

Next, John and Liz grind the exterior and interior surfaces smooth with heavy grit sandpaper at graduated levels of abrasiveness. This removes most of the sand texture, and leaves a smooth matte surface on the casting. The interior edges of the handles are then refined with a small handheld belt grinder. After the finishing work is completed, the skillets are ready for seasoning. All up, it takes about a week to produce a pan.

Leaving as small a carbon footprint as possible has always been important to John and Liz. In the beginning, they sidestepped coal by using a homemade furnace that ran on waste vegetable oil – the Skilletron. In 2016 they upgraded to an electric tilt furnace, and now run their entire workshop with an on-site solar array. They use only recycled iron as source material for their castings, and they also crush up and reuse the sand they use in their moulds, ensuring there are no waste byproducts resulting from their making process.

'Our skillets have all the benefits of cast iron, but are easier to hold, and the handle remains cool to the touch. I love the way our small-batch manufacturing allows us to connect with each and every one of our customers. People write and tell us what they've been cooking; it's really inspiring,' says John.

"

I love the way our small-batch manufacturing allows us to connect with each and every one of our customers. People write and tell us what they've been cooking; it's really inspiring.

CARING FOR YOUR CAST-IRON FRYING PAN

A well-seasoned and cared for pan has a surface as slippery as a fish's belly, and a taste that improves with age. Here are some tips to keep your pan shipshape:

- Pre-warm the pan and oil before adding ingredients. This helps to prevent sticking, especially in the first few uses.
- Avoid cooking acidic foods such as tomato sauce in your skillet. They degrade the seasoning and can absorb a metallic taste.
- Do not use olive oil for seasoning or re-seasoning. It doesn't form an effective base layer, as it comes away from the iron a little too easily.
- Never let your skillet stay wet for long, as the iron can rust. If this happens, gently scrub or sand the rust off with a scrub sponge and re-season the pan.
- If your pan's cooking surface becomes sticky, try cooking with a little more oil than usual. If it doesn't improve, it's time to re-season your skillet.
- Never wash your pan with soap, or in soapy water.

HOW TO

CLEAN YOUR CAST-IRON FRYING PAN

MATERIALS

- Warm water
- Coarse salt
- Olive oil

TOOLS

- Dish brush or scrub sponge
- Copper scrub pad
- Microfibre cloth

METHOD

1. For standard cleaning, rinse the pan under warm water, then brush it out with a dish brush or scrub sponge.
2. For more intensive cleaning, rub the dry surface with coarse salt, or use a copper scrub pad under warm water. Copper is softer than iron, so it won't scratch the surface of your pan. Dry the pan straight away.
3. Use the microfibre cloth to rub a tiny amount of olive oil onto the pan's surface to further protect the iron from moisture.

Step 3

HOW TO

RE-SEASON YOUR FRYING PAN

MATERIALS

- Flaxseed oil, lard or oil of your choice

TOOLS

- Oven preheated to 95°C (200°F)
- Microfibre cloth

METHOD

1. Warm the pan in the oven for 10 minutes.
2. Remove the pan and place it on a heatproof surface, then increase the oven temperature to 260°C (500°F) if using flaxseed oil, or 200°C (400°F) for other oils and lard.
3. Return to your pan. Use the microfibre cloth to rub oil onto its surface, then rub almost all of it off again. You want to leave a very thin layer of oil behind; if it's too drippy or thick, it can flake off.
4. When the oven has warmed to the correct temperature for the type of oil you are using, return the pan to the oven. After an hour, turn the oven off but leave the pan inside until it is cool enough to touch. It shouldn't be sticky – if it is, that means it needs more oven time. Repeat the process in 15-minute increments until it's no longer sticky.
5. While the pan is still warm, repeat steps 2 and 3 up to four times.

532
531
530

GRAFA

TUBE GARDEN TROWEL

Coppersmithing

Copper is about as perfect for making gardening tools as it gets. It doesn't rust or leach toxins into the soil; its properties may even promote growth and nutrient uptake. This trowel, the Tube, is designed for digging in established garden beds, courtyard and balcony gardens, and for indoor gardening and potting. It is ideal for planting seedlings and bulbs, and removing weeds. Designed with durability in mind, it's made from a single piece of reclaimed copper, and handles difficult or clayish soils with aplomb. Its maker, a qualified fitter and turner and self-taught coppersmith, says that copper is malleable enough to shape into interesting and organic curves but strong enough to last a lifetime in the garden. It is also infinitely recyclable; 50 per cent of copper produced today is made from recycled copper. Now that really puts a sheen on things.

MEASUREMENTS	Length: 260 mm (10 in); weight: 190 g (7 oz); diameter (handle): 19 mm (1 in)
MATERIALS	Reclaimed copper
KEY TOOLS	Forming tools, hammers, metal linishers, snips, stamps
KEY MACHINES	Cold saw, metal linisher
TIME TO MAKE	20 minutes
LIFESPAN	Forever

TRAVIS BLANDFORD — Coppersmith [Melbourne, Australia]

It was Travis' love of gardening that led him to copper. After he completed a permaculture course, he decided to apply his trade skills to the world of gardening, and began reading about copper's potential for benefiting soil. Around the same time, he noticed that copper tubing – usually recycled off-cuts from building sites around town – was readily available at various metal recyclers.

It got him thinking. In 2011 he sourced some copper and designed and made his first prototype: the Tube. More designs followed, and now, under the name Grafa, he makes gardening tools made from copper, bronze and wood.

The Tube is made from a single piece of copper, which Travis cuts open to form the handle and the shape of the blade. It's the simplest of his designs; even so, there are more than ten steps involved in making one.

Travis begins by donning his safety gear – gloves, glasses and sometimes a face shield – then gets to work cutting the copper to size with various snips and saws. Then he opens and shapes the tube with various purpose-built forming tools – some of which he has made himself – to create the Tube's curved shape.

He cuts the tip or blade of the Tube using tin snips, then stamps the blade with the Grafa name and logo using a metal stamp. He uses a metal linisher to clear the burrs from the edges of the blade, then various grades of brushes and steel wool to create a uniform finish on the surface of the tool. Next, he puts the end cap onto the end of the tool with a hammer.

'The final step is coating the surface of the tool with a linseed-based clear coat to keep the tool nice and shiny before it hits the soil. Then I place the Tube in a drying cabinet to dry overnight, before packaging it up, ready to ship.'

Copper was one of the first metals mined and turned to tool and decorative use by humankind. In fact, there's an entire age named for its use, the Copper Age, which historians generally agree existed from 5000 BCE to 4000 BCE, when the Bronze Age began. A copper pendant dating to 8700 BCE has been found in northern Iraq, while the archaeological site of Belovode, on the Rudnik mountain in Serbia, contains the world's oldest evidence of copper smelting.

In our time, it was Austrian forester and inventor Viktor Schauberger (1885–1958) who made the connection between the use of copper tools and implements in soil, and increases in plant growth and yield. He conducted field trials with iron and copper, and observed that the crops cultivated with copper ploughshares had larger, healthier yields and fewer pests than those cultivated with iron. He began to advocate for the use of copper in soil.

Travis works in a 32 sq m (344 sq ft) workshop in a one-time timber joinery factory, leasing the space from someone who runs a foundry. Travis has a central space with a skylight, and has added a mezzanine for his music, and a storage area.

'At the moment I pretty much work alone; sometimes that can be a bit hard going, but my workshop is surrounded by other designers and artists so there is always someone to chat to if I need a break. I work in silence when I'm working with machinery, but when I'm assembling tools, I like to listen to fast rock.'

When things get busy, Travis's partner Harriet (also Grafa's business, marketing and production coordinator), a friend, or his father, who is a retired boilermaker, come in to help out.

'The best things about copper are that it's malleable enough to work with easily and that, with correct methods, you can make tools that will last for a long time,' says Travis.

'I take great satisfaction in completing an order and knowing the tools will live on in a garden somewhere in the world, and help that garden live on too.'

CARING FOR YOUR TUBE

Always put your Tube away in the garden shed after use, taking care to keep it away from harsh garden chemicals or household cleaners. To clean the tool itself, simply wipe it down with a damp rag. A natural patina will emerge in the copper over time. This patina will come and go with use in the soil; more use means less patina. If you want to remove the patina, gently rub it with fine-grade stainless steel wool. You can also treat the surface with a high-quality metal polish, though this is purely for aesthetic purposes.

"

I take great satisfaction in completing an order and knowing the tools will live on in a garden somewhere in the world, and help that garden live on too.

HOW TO

MAKE A COPPER SNAIL COIL

Plants and soil love copper, but slugs and snails? They hate it. These easy to make copper snail coils will protect your seedlings from slimy garden pests.

MATERIALS

- 2–2.5 mm (1⁄10 in) solid core copper wire (available from metal recyclers and electrical wholesalers)

TOOLS

- Pliers
- Hammer (optional)
- Metal file
- Bench vice
- Rigid cylindrical object 50 mm (2 in) in diameter – a tool handle will suit

METHOD

1. You'll need a decent bench space to work on for this project; 1 sq m (11 sq ft) is ideal.
2. Use the pliers to cut the wire to lengths approximately 30–40 cm (12–16 in) long.
3. Start bending the wire 10 cm (4 in) along and wrap it around a cylindrical object a couple of times to create a coil 5 cm (2 in) in diameter. Leave enough excess wire so there is a 4-cm (1½ in) overhang at the end. This will take a bit of trial and error to get right in the beginning.
4. Bend the original 10-cm (4 in) length sharply, so that it intersects the diameter of the copper circle you've created.
5. Use a hammer to flatten the two ends (this step is optional, but it makes handling safer).
6. Use the metal file to clean up the ends so there are no burrs.
7. To use the coil in the garden, bend the 10-cm (4 in) section out at right angles so it's perpendicular to the circle and will pierce the soil. This fixes it in the soil and means the coil can sit around a seedling to protect it from snails.

THE GALILEO GLOBE

Globe making

Apparently flat Earth theories are still a thing. There's even an official society – though its members clearly haven't had the pleasure of poring over (and around) a spherical globe like this one. Meet The Galileo, made by Peter Bellerby and a committed team of round Earthers in a light-filled atelier in North London. Inspired by the famous Blaeu globe of 1599, The Galileo is politically and demographically current. It has brass meridians and a solid oak base, and can be moved 180 degrees. Globe making is nothing new – a Greek grammarian, Crates, is credited with making the first globe in around 150 BCE, and they were a must in any well-to-do household for many years. But globes of this calibre have been on the endangered list for quite some time – until Peter came along.

MEASUREMENTS	Globe diameter: 80 cm (31½ in); table diameter: 120 cm (47 in); total height: 135 cm (53 in); weight: 50 kg (110 lb).
MATERIALS	Globe: plaster of Paris, counterbalance weights, glue, paper, watercolour, ink, resin, varnish; Meridian: brass; Stand: wood of choice (this one is oak)
KEY TOOLS	Paintbrush, pencil
KEY MACHINES	Band saw, belt sander, chop saw, hand sander, lathe, mitre saw
TIME TO MAKE	4 months
LIFESPAN	Heirloom quality

PETER BELLERBY — Globemaker [London, United Kingdom]

All Peter Bellerby wanted to do was buy his father a top-quality globe for his eightieth birthday. In the end, faced with the choice between a substandard factory model or an outdated, fragile antique, he decided to make his own. Almost two years and a lot of discarded materials later, he'd finished a globe ... and decided to start a business.

These days, Peter and his 15-strong team are Bellerby & Co Globemakers, and they have the process down to a fine art. They make around 350 globes a year, from desktop models (which can be moved 360 degrees) to ones like The Galileo, which practically warrants its own quarters.

They begin by making the sphere. The globes are cast in two halves – the northern and southern hemispheres – from plaster of Paris in resin moulds. Counterbalance weights are placed inside the two halves to ensure the globe will spin smoothly and come to a gentle, controlled halt, and then the halves are glued together.

The spheres weren't always formed this way. When Peter started working on his very first globe, he tried to make the sphere himself, using plaster of Paris and mathematical instruments.

'I learned very quickly that the margin for error is tiny and so much can go wrong. And it did! Every time I messed up a measurement, I was multiplying that mistake by Pi. That meant that by the time I'd finished the shape, it was way off.'

When the sphere is ready, the paper maps are prepared and printed. They're created on software that translates a regular map into triangular sections called gores.

Any illustrations the client has requested are hand drawn, then scanned and placed onto the cartography in light outline. The gores are printed, then cut by hand, and the first of many watercolour layers for the ocean are painted on by hand.

After the gores have dried, the most difficult and painstaking part of the entire process begins – applying them to the sphere.

'The goring process is long and requires complete concentration. The paper is wet, so it will take the shape of the globe properly, but that makes it very fragile,' Peter explains.

'It's a case of learning how to use your hands in a very delicate manner. Every movement has to be in slow motion for fear of damaging the paper. At times there will be five of us papering away on a big project and there won't be a word shared between us.'

When the goring of the globe is finished and checked over, the artist comes back and paints any custom illustrations, and more layers of pigment in the ocean and shade around the coastlines. Peter uses Sennelier watercolours along with hand-mixed pigments and a range of paint brushes. The most he'll give away is to say that 'Squirrel mop brushes are essential'.

The finishes of many antique globes suffer from blackening, because of the acidity in glues that were used by past makers. Peter takes care to use a pH-neutral glue, and coats the globes in a UV protective varnish, to ensure the colours stay fast.

When it's dry, the globe is carefully attached to its base, which is made on site variously using a band saw, belt sander, chop saw, hand sander, lathe and mitre saw, and to its brass meridian. The meridian is hand cast by offsite metalworkers then engraved and finished by hand in the Bellerby & Co studio – usually in downtime, while the freshly applied gores or layers of paint are drying on the spheres.

Peter has tested the robustness and integrity of his globes by subjecting them to all sorts of tests – even going so far as to immerse one in a bucket of water for days on end. He's proud to say the hardy globe retained its integrity.

'The driver for me to first make a globe was that so many have fallen apart due to poor materials or methods. We use only the best pigments, inks, glues, resins and varnishes, all of which help to maintain the longevity of our globes. I want them to be passed down through families for generations.'

CARING FOR YOUR GLOBE

Do not leave your globe in direct sunlight or right beside a fire or heating unit for long periods of time. The globe itself can be wiped with a clean, dry cloth from time to time, and the wooden stand can be cared for as you would other wooden furniture.

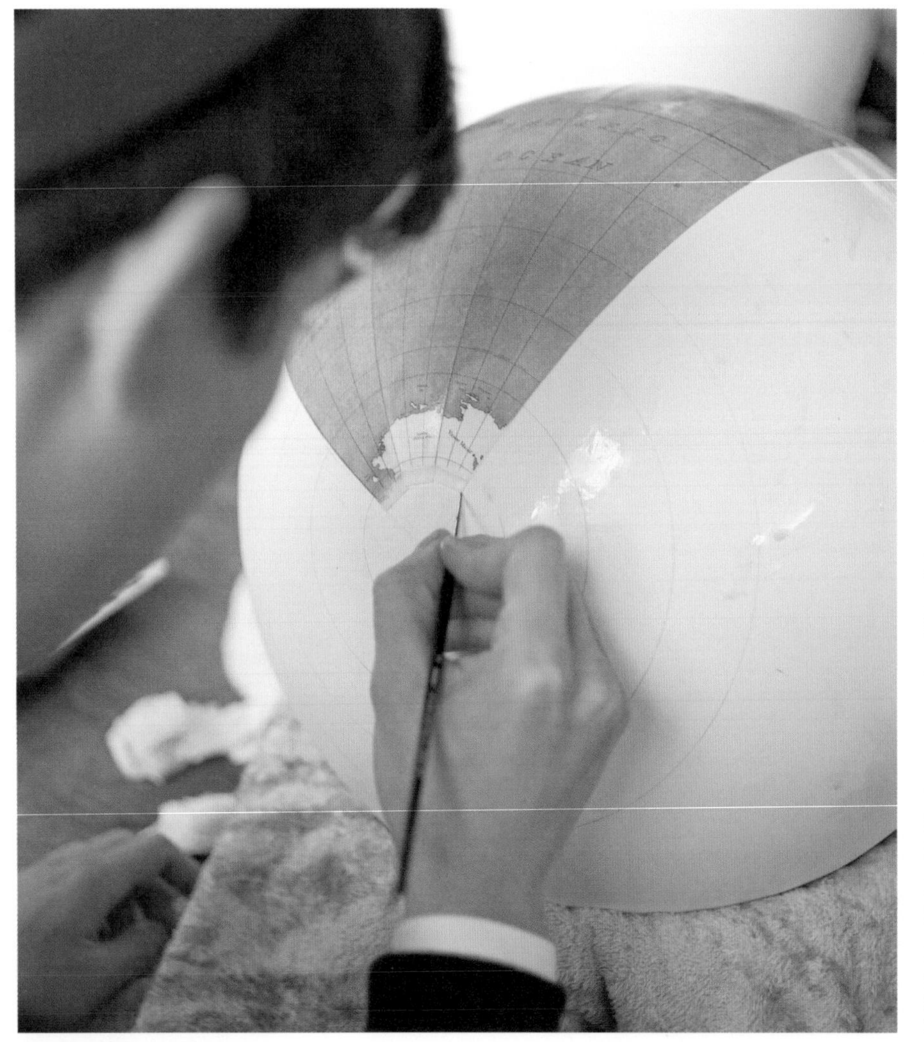

“

It's a case of learning how to use your hands in a very delicate manner ... At times there will be five of us papering away on a big project and there won't be a word shared between us.

HOW TO

MAKE A GLOBE CHANDELIER

Peter wouldn't be drawn on any of the secrets behind his stunning globes, but he did share this cheeky project, which repurposes second-rate plastic globe lamps into a funky chandelier. One hangs ironically in the Bellerby & Co studio.

MATERIALS

- 6–8 old or second-hand plastic globe lamps, ideally of varied sizes
- 1 m (39 in) long stainless steel rods – one for each globe
- 10.2 cm (4 in) diameter flat washers – two per globe
- Nuts to fit on the stainless steel rods – two for each globe
- Lightbulbs (one for each globe) with the right fittings – bayonet caps or Edison screw caps
- 6- or 9-plug multiplug with surge protection
- Extension lead

TOOLS

- Cable ties or gaffer tape
- Spanner (wrench)

METHOD

1. Unscrew each sphere from its base. Take care to retain the light fitting and electrical cable that comes with each lamp, but recycle the plastic bases.
2. Take a length of stainless steel rod and thread it through one of the globes.
3. Take a washer and thread it onto the base of the globe, then add a nut and screw it on tight using the spanner (wrench).
4. Put the globe carefully aside, then repeat this process with the remaining globes, grouping them together on the back of a couch or similar as you go.
5. Put a new lightbulb into a light fitting on one of the spheres, then gently lower the bulb and electrical cable down through the hole in the top of a globe. Secure a washer and nut on top of the sphere, then lay the remaining cable and plug over the back of the couch. Repeat for all the globes.
6. Insert each plug into the multiplug, taking care not to get the cables tangled.
7. Gather the wires and arrange the globes as you want them to hang – some higher and some lower, until they sit well.
8. Tie them together using the cable ties or gaffer tape.
9. Hang them from a suitable spot – a beam? A hook? A light fitting on the ceiling? You may need a ladder and a spare pair of hands for this stage.

NOTE

This project assumes that you're not an electrician, and therefore shouldn't be messing around with wires. But if you are, you'll want to remove all but one of the switches, rewire the bulbs and cables through the one switch, and ditch the multiplug.

Lee Valley

PRIMROSE ELECTRIC GUITAR

Luthiery

Learning to play guitar? Secret dream of pretty much everyone who doesn't already know how. Learning to make a guitar? Epic. No-one knows exactly where or when the first guitars were made, but they sprung from medieval Spain. A descendant of the lute, today's guitars are either acoustic or electric, and steel or nylon stringed. Of course, descriptions like these don't do justice to the guitar's potential to inspire musical greatness, or emit chords of such sonic beauty that they'll make you weep. This guitar has been crafted from a chorus of materials including abalone, deer antler, mahogany, birch, and maple, and was made by one of the founders of the Canadian School of Lutherie, Mitch MacDonald.

MEASUREMENTS	Total length: 99 cm (39 in); scale length: 64.8 cm (25½ in); widest point on body: 33 cm (13 in); depth at centre: 5 cm (2 in); depth at outside edge: 4½ cm (1.8 in)
MATERIALS	Abalone, deer antler, high gloss pre-catalysed lacquer mahogany, Macassar ebony, polymerised tung oil, smoked birch, smoked maple (guitar), baby grand bridge, locking tuners, knobs, pickup rings, strap buttons, truss rod cover and pick guard (hardware), humbucking pickups, volume, three-way switch, tone (electrics)
KEY TOOLS	Drills, files, hand planes, hand saws, level, palm sanders
KEY MACHINES	Band saw, drill press, jointer, routers, router table, sanding machines, table saw, thickness planer, thickness sander
TIME TO MAKE	4 days
LIFESPAN	Heirloom quality

MITCH MACDONALD — Luthier [Prince Edward Island, Canada]

Holding a guitar is one of Mitch MacDonald's earliest memories. He remembers the guitar being huge, so he must have been very small. His father had bought it for Mitch's mother as a gift, but she never played it. So Mitch dug it out of the closet and dragged it up onto a chair, laying the foundation for a lifelong love of the guitar.

He didn't actually learn to play until he was a teenager. But when he did, he took the art seriously, practising for at least 30 minutes a day – a discipline he continues even now.

Mitch learned his woodworking skills at a two-year-long Heritage Carpentry program in Nova Scotia. An apprenticeship with luthier George Riszanyi followed; Mitch later co-founded the Canadian School of Lutherie with George, as well as Jeremy Nicks of Lethean Guitars in Toronto.

So far, Mitch has built nearly 30 guitars, each unique. Once he has decided on the style of guitar he wants to build, he selects the wood and uses a table saw, band saw, planer, jointer and thickness sander to mill it to the appropriate sizes. He glues the body together and leaves it to dry for a day or two, then uses a router to cut out the profile of the guitar's shape or style.

'Mahogany is a beautiful wood that is easy to carve and cut and can be fairly light. The smoked maple top gives the guitar body a vintage look of aged lacquer without adding stain. In luthiery, mahogany is renowned for having what we call a "warm tone", which is hard to articulate until you've heard it in person.'

Mitch routs all the necessary pockets and cavities in the body, and carves any contours or arm bevels. Next, it's time for sanding. This part of the process takes around 3 hours. Mitch starts with 80 grit sandpaper and works his way to a finish of 800 grit. He then applies a finish to the body, and gets to work on the fretboard.

'The Primrose's fretboard is made from Macassar ebony; it has a dramatic, striped appearance and is really eye-catching. I installed abalone fret markers and a deer antler nut. I buy the abalone in

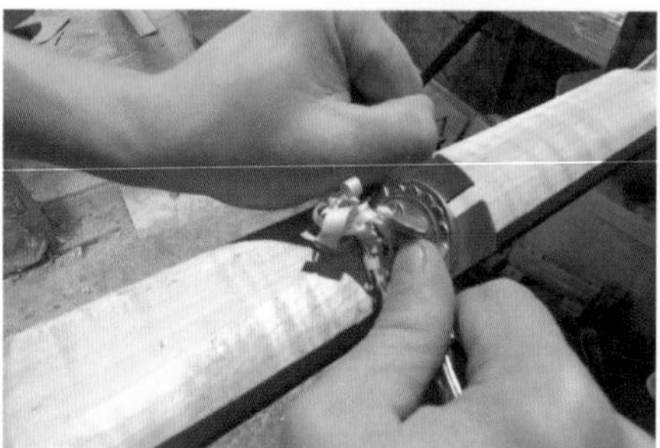

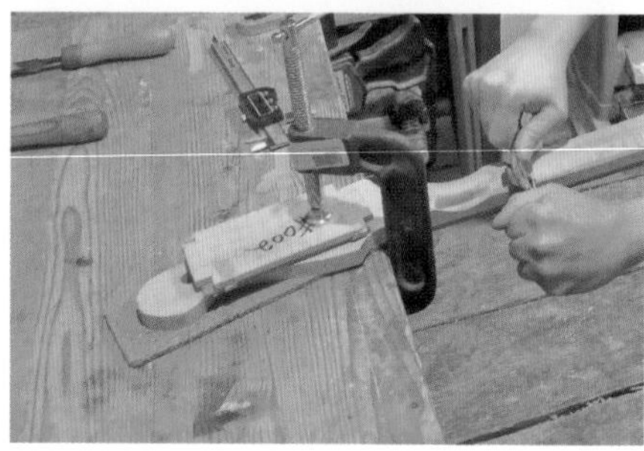

“

I love the ambience of a woodworking shop; the smells, the rough timber sitting on the shelf begging to be used for a project ... It is a sacred place for me.

pre-cut dots and glue them in place, then file and sand them down to be flush with the fretboard's surface. The antler nut, I cut from larger pieces on the band saw. I have a collection of shed moose and deer antlers that I choose from.'

Mitch routs a channel in the neck and installs a truss rod, then glues the fretboard to the neck. When it's dry, he uses the router to cut the profile for the neck. Next, Mitch installs the frets, files them with a bevelling file, sands them, levels them, rounds them with a crowning file, then polishes the neck.

'The crowning leaves file marks and the sanding leaves the frets scuffed up and dull looking. Polishing makes the frets look great, and hardens the metal by compressing the material. It's achieved simply; by sanding. I start with 320 grit and work my way right up to 12,000 grit, then follow up with a metal polish.'

He carves the neck to the desired profile, drills holes for tuners, and thicknesses the headstock – the very top of the guitar where the tuners that hold the strings are installed– by carefully shaving it down on a sanding jig that affixes to a spindle sander.

He sands the neck and applies the finish, then bolts the neck to the neck pocket on the guitar, and installs all the hardware and electronics. Finally, Mitch plays the guitar and makes final adjustments.

'I love the ambience of a woodworking shop; the smells, the rough timber sitting on the shelf begging to be used for a project. I get a feeling of accomplishment and fulfilment that I'm sure many people feel when they are creating something. It is a sacred place for me.'

CARING FOR YOUR GUITAR

Guitars change over time. They get played until the frets are worn. They warp slightly in changes of season, temperature or location; the neck bends under string tension and they become hard to tune. Environmentally, humidity is your biggest foe. Too much or too little is not good for the instrument. Too dry and you'll get cracks and lifting bridges; too damp and the guitar will struggle to stay in tune and may sound 'muddy'.

When the strings start to sound dull, it's time to change them. If you play regularly, they'll need changing every one to three months. While changing the strings it is also a good time to clean the entire guitar using a guitar-specific polish, and to wipe a thin layer of lemon oil onto the fretboard. Let the oil sit for 5 minutes, then buff it away with a clean cloth. Between playing, keep your guitar in its case. If it's going to sit unplayed for a long time, or if you're travelling with it, slacken the strings. Lastly, if you notice a crack forming or a glue seam letting go, you have a guitar emergency. Remove all tension and take it to a luthier, stat.

HOW TO

MAKE A SIDE BENDER AND BEND A SET OF GUITAR SIDES

Normally, we shape wood via sanding, planing, cutting and so on. But this simple method involves manipulating the shape of the wood by applying heat and moisture, bending it by hand, then clamping it in a jig or form and letting it cool. Mitch does this on every acoustic guitar he makes.

MATERIALS

- Aluminium can – approximately 5 cm (2 in) diameter and as long as you can find
- Piece of plywood approximately 10.2 × 25.4 cm (4 × 10 in)
- 3 bolts, wing nuts and washers
- 15.2 cm (6 in) long and 2 mm (⅛ in) thick spring steel (the length determined by the length of the can)
- 1–3 self-tapping screws
- Metal zip ties or pipe clamps
- Small candles (tea lights are perfect)
- 2 pieces of wood measuring 91.4 cm (36 in) long × 12.7 cm (5 in) wide × 2.4 mm (⁹⁄₁₀ in) thick for the guitar sides – Mitch recommends rosewood for beginners

TOOLS

- Tin snips
- Band saw (or handsaw and vice)
- Pencil
- Vice
- Drill bit to match diameter of bolts
- Hand drill
- Permanent marker
- Screwdriver with socket attachment
- Hammer
- Nail set
- Spray bottle filled with water
- Jig or form or mould for your guitar

SAFETY GEAR

- Safety gloves

METHOD

1. Put on your safety gloves. This is important; the aluminium can and the spring steel are very sharp.
2. You need to have one end of the can open; the other solid. If the aluminium can isn't empty, empty it (and make good use of the contents!) and remove one end by hand, if the can allows, or by using the tin snips.
3. Use the band saw (or handsaw and vice) to cut a piece of plywood big enough that, once the can is attached, it can be clamped into a vice without getting in the way. Mitch recommends 10.2 × 25.4 cm (4 × 10 in), but you may need to adjust the measurements for your project.
4. Place the closed side of the can on the plywood, then use a pencil to trace its diameter onto the plywood.
5. Make sure the drill bit matches the size of your bolts (if the bolt is ¼ in then you need a ¼ in drill bit), then drill 2–3 holes, equally spaced 6.4–12.7 mm (¼–½ in) in from the edge of the traced circle. These holes will be for bolting the can to the ply. Drill one hole in the middle of the traced circle to allow for airflow.
6. Line the can back up with the marking you traced, and mark the traced holes on the can with a marker.
7. Now drill holes in the can where the marks are.
8. Use a screwdriver with a socket attachment to fasten the can to the plywood base with bolts. Set it up so you're putting the wing nuts on the outside of the can; this makes it easier to tighten the wing nuts.
9. Take the spring steel and wrap it around the can once, and mark where it overlaps. Keep the steel as tight as possible, without bending the can. Use the tin snips to cut the spring steel, leaving a 2.5-cm (1 in) overlap.
10. Now use the nail set to punch three holes half the width of the overlap in from the edge of the spring steel. Mitch's overlap is normally 2.5 cm (1 in), and his punching position 1.25 cm (½ in).
11. Wrap another length of steel around the can and draw it tight again.
12. Mark three additional holes in the steel with a marker and punch them out.

HOW TO

MAKE A SIDE BENDER AND BEND A SET OF GUITAR SIDES

13. Wrap the steel around the can again and match up the holes, then use a self-tapping screw to fasten the steel to the can.
14. Add a pipe clamp to both ends of the cylinder and tighten.

Bending the sides of a guitar

1. You're now ready to insert the candles into the bender (Mitch uses three tea lights), light them, and start bending. You may have to play around with the spacing of the candles inside the bender; one might keep going out until you get the spacing right.
2. Once you have them right, the unit will heat up extremely quickly; you'll be ready to start bending within a couple of minutes. You can test it by putting a drop of water on the bender; once the water starts to sizzle, it's ready.
3. Take one guitar side and spray it with water from the spray bottle. The water will turn into steam, which softens the wood and helps it bend. It also keeps the wood from burning. Keep the bottle within reach, as you'll need to spray the side again during the bending process.
4. Now hold the sprayed guitar side perpendicular to the cylinder. Lay it flat on the cylinder and press down firmly, but not too hard, or the wood will crack. As it warms you'll feel the wood start to flex; this means it's working. Move and bend the wood as it heats.
5. As the water evaporates off the wood, the wood will stop flexing; this is a good time to take the side to the jig and check to see how your bending is going.
6. Spray the wood again, then take it back to the bender and continue to roll and bend the wood. Repeat this process until the side has the necessary curve; different woods take different amounts of time to bend. A rosewood side takes Mitch about five minutes to bend.
7. Once you're confident you've bent the side to the shape of the guitar jig, clamp it in the jig to preserve the shape.
8. Bend the second side in the same manner.
9. Blow out the candles once you have finished bending.

Step 1

Step 3

Step 4

Step 6

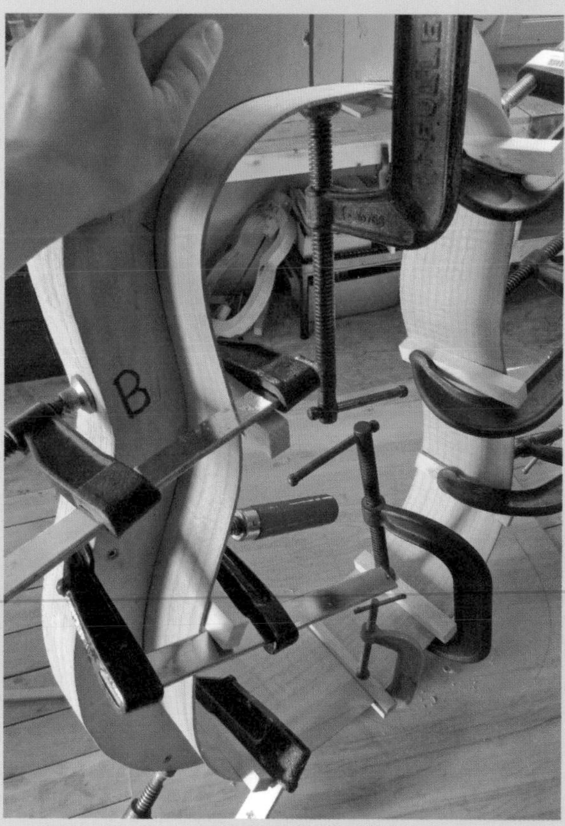

Step 7

HOW TO TIE A TIE
Otis James
THE BOW TIE
THE FOUR-IN-HAND KNOT
OTIS JAMES
-NASHVILLE-

WAXED COTTON CAMPER CAP

Millinery

Waxed or oiled cloth has been around for centuries. Initially used for making sails, today it's used mostly to make clothing for hunters and motorcyclists. This six-panel ball cap is made from waxed cotton canvas. It will last for years – maybe even decades – and its appearance will only improve with age (and wear). Hand made by a modern-day cap maker in the basement of a mid-century house in Oak Ridge, Tennessee, it is tough and simple. There are no unnecessary embellishments, linings or stitching. It's designed to be crumpled up in your bag or stuffed in your pocket, and taken out hours later to be worn. It's a hat for everyone – wear it in the city, in the country, in the bath, or in bed. It'll stand up to anything you throw at it, and then some.

MEASUREMENTS	Circumference: 55.9–63.5 cm (22–25 in)
MATERIALS	Cotton seersucker, waxed cotton cloth, heavyweight duck canvas
KEY TOOLS	Scissors, rotary cutter
KEY MACHINES	Cylinder arm needle-feed sewing machine, needle-feed sewing machine, straight-stitch sewing machine, walking foot machine
TIME TO MAKE	2 hours +
LIFESPAN	10 years +

OTIS JAMES — Cap maker [Tennessee, USA]

Otis James started out on his own in 2009, making neckties out of a shed in his backyard. By 2013, he had a studio, five employees and a retail space. He looked successful, but he didn't feel it. He felt ... distracted. He was busy managing a business and people who were busy making the things he had once made himself, but no longer had the time to make.

'I wasn't in a position to just walk away and start over, so I began to slowly dismantle what had been built up. In early 2017 I started getting back to the roots of my endeavour. I have a private studio in the basement of a house in Oak Ridge. It's just me. It's modest, by many measures, but it is exactly what I desire,' says Otis.

These days, Otis is back to doing what he loves – making caps and bow ties and communicating directly with his customers. Caps are the mainstay of his business, and the waxed cotton Camper is one of his most popular designs.

'The Campers were designed to be a distillation of a great everyday cap. I love simplicity. So the waxed Camper has only as many stitches as I thought necessary to make a good-looking, durable cap.'

Each Camper takes Otis around 2 hours to make. He begins by cutting the materials – waxed cotton cloth for the shell, cotton seersucker for the inner band, and a heavyweight duck canvas for the stiffener in the peak – using a manual rotary cutter. There are six panels on the crown, an outer band, two pieces for the peak cover, the canvas peak stiffener, an inner band, interfacing for the inner band, and a cotton backing to the inner band.

Next, Otis trims the outer and inner bands to the appropriate length for the intended size. He makes notches with scissors for attaching the peak, and stamps his name on a cotton ribbon label. Finally, he sews everything together.

'I do the sewing on four different machines. I have an old Pfaff industrial needle-feed for the inner band and backing that is set up to handle lighter weight materials. I sew the crown on an old industrial Singer straight-stitch. I sew the peak on a heavy duty

CARING FOR YOUR WAXED COTTON CAMPER

The simplest way to make a cap last a long time is to rewax the cap when it starts to get really dry. The canvas itself is very durable, but the wax offers a layer of protection that keeps the cloth from too much abrasion or exposure to the elements. When the cap gets dirty, brush it clean or wipe it with a damp cloth. Don't let the cap stay damp, and don't use soap. And keep it away from animals that like to chew hats. They'll love it as much as you do.

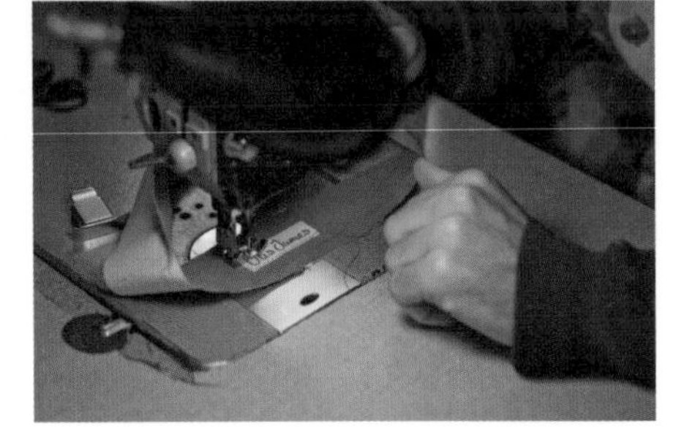

Consew walking foot machine to make the seven stitch lines that give the peak stability and some rigidity. Then I attach the peak to the crown and the inner band on another old heavy duty Singer, a cylinder arm, needle-feed machine. Then the final seam, which secures the inner band to the crown, I do on the walking foot machine.'

Otis prefers to use old industrial equipment in his workshop.

'It's just much better made. I most enjoy sitting down at the sewing machine to create the crown of the cap. That is when the cap really starts to take shape. And I do that part on my favourite machine. It was the first industrial machine I bought back in 2009, so it's sort of like sitting down with a best friend. We've been through a lot together.'

Otis made his first cap in 2009, aged 25. 'It was a replica of the first cap I ever bought. That was a grey tweed ivy cap that I bought at age 14. Eleven years later, I took it apart to make a pattern from it and made a similar cap in a grey tweed.'

He's committed to making a quality product with strong attention to detail. Each item he makes is crafted with love and care. Every detail is hand finished, from the stamps and painting on the labels to the buttonholes and slip-stitches on each tie.

'Sure, it takes longer this way, but quality and character are the ideals, not mass-production. I suppose I'm old-fashioned that way. I just truly believe there is something to be said for knowing exactly where something was made, and by whom.'

"

... quality and character are the ideals, not mass-production. I suppose I'm old-fashioned that way. I just truly believe there is something to be said for knowing exactly where something was made, and by whom.

HOW TO

WAX FABRIC

MATERIALS

- Fabric or garment to be waxed (usually cotton, but could be hemp or linen)
- Fabric wax (in a bar or a tin)

TOOLS

- Lint roller or masking tape
- Hair dryer or heat gun

METHOD

1. First, clean your fabric. Remove any debris with a lint roller or a piece of masking tape. Be aware that whatever is left on the fabric will be trapped in the wax and remain visible, so do your darnedest to get it all off! If the fabric has been waxed before and is dirty, hand wash it in cold water with a rag or brush. Do not use soap or detergent.
2. Prepare the wax. On a warm spring or summer day, the wax should already be soft enough to apply, but if you're waxing in a colder climate, you'll need to soften the wax. This can be done with a little bit of heat from a hair dryer, by sitting the wax on a heater vent, or otherwise applying warmth to it.
3. Apply the wax. If the wax comes in a bar form, rub the bar all over the fabric using broad, overlapping strokes. The wax will soften with the friction and become easier to spread as you go. If the wax comes in a tin, apply it liberally with your fingers, again in broad strokes. If waxing a garment, use your fingers to work the wax into seams and around attachments such as handles and zippers. Try to get even coverage throughout.
4. Penetrate the cloth. Use a hair dryer or heat gun on medium to high heat to soften the wax further, allowing it to penetrate into the weave of the fabric. Use your fingers as needed to work it into seams, crevices or difficult areas. This will give the waxed finish more longevity.
5. Let the wax dry. Give it a day to set in the fabric, then wear it to your heart's (or other relevant body part's) content.

Step 3

Step 4

Step 5

FIRE CIDER

Fermenting

Fire cider is a potent folk remedy that promises to boost your immune system and ward off colds, flus, sinus congestion and other lurgies like a medieval swordsperson. It also aids digestion, increases circulation and helps to keep you in tiptop physical condition. It's hot, pungent, sour and sweet, and as its alternative names, plague tonic, fire water and master tonic, suggest, it's not for the faint of heart. Fire cider is a fermented (albeit non-traditionally fermented) brew, which means it converts carbohydrates to alcohol or organic acids using microorganisms – yeasts or bacteria – under anaerobic conditions. It's incredibly easy to make, and most committed food fermenters, like writer, photographer, blogger and mama extraordinaire Brydie Piaf, who has been fermenting since 2010, have fire cider on stand-by in their pantries year round.

MEASUREMENTS	1 litre (34 fl oz/4 cups)
MATERIALS	Chilli, garlic, ginger, honey, horseradish, onion, raw unfiltered apple cider vinegar, rosemary, thyme, turmeric root (all organic, if possible)
TOOLS	Bottle, bowl, chef's knife, chopping board, funnel, strainer, watertight glass jar
MACHINES	Nil
TIME TO MAKE	14–28 days
LIFESPAN	Forever

BRYDIE PIAF — Food fermenter and writer [Newcastle, Australia]

Brydie Piaf first made fire cider in 2014, after learning about it through a friend. She brews it several times a year, and drinks it most in winter, when there is often an increase in lurgies brought home by her three busy children. Brydie's husband and youngest daughter happily drink the fire cider, but her older two kids? Not a chance.

For a potion so potent, fire cider is incredibly easy to make, and there are no hard and fast rules when it comes to the ingredients. Once the basics of chilli, garlic, ginger, onion and turmeric root go into the jar, you can add your own nutritional favourites. You like horseradish? Add horseradish. And so on.

The way the ingredients are prepped is up to the maker, too. Brydie likes to chop hers into roughly even sized pieces. She begins with the core ingredients, then adds dashes of her chosen extras: horseradish, thyme and rosemary. Next a big lug of apple cider vinegar (which she also makes herself) is poured into the jar.

The jar is then sealed tight and put out of direct sunlight on a far corner of the kitchen bench. Brydie gives it a gentle swish once or twice a day. After a month, she opens the jar, strains it into a bowl, mixes in a generous spoonful of raw honey and decants that into a bottle. The lid goes on and it's put in the pantry, ready for drinking in small, shot glass-style doses.

'I feel a lot healthier with fermented foods in my diet. They weren't something I grew up with, but backyard vegetables and freshly baked bread was,' says Brydie, who blogs at cityhippyfarmgirl.com.

'I spent a lot of time in the kitchen as a kid, either stirring, rolling or simply watching my mother make everything from scratch. Things would be preserved, pastry would be rolled out and vegetables would be turned into dinner. I learned by osmosis really, simply being there and helping while talking or doing homework.'

Large, wide-mouthed glass jars and bottles to decant into are a must, as is good salt. 'You can do a fair whack with just those things,' says Bridie, whose rows of fermenting foods and tonics and jars plump with preserves take pride of place in the kitchen.

'They're a little part of the season I've managed to capture and they make meal times all the more delicious. I love not having to rely on shops for much more than just bulk items, and it's incredibly satisfying seeing my partner and kids get obvious enjoyment from eating the way we do.

'I'd been making bread for a couple of years and wanted to take it a step further, so I created a sourdough starter and haven't looked back. I was really happy with those first experimental baking results and still regularly bake sourdough each week. It opened up the fermenting door. What else could I ferment?'

Fire cider is one of Brydie's favourites, but she's also tried her hand at ginger beer, kombucha, mead, pickles, sauerkraut, and apple cider vinegar. Mostly things work out, but Brydie's methods definitely leave plenty of room for flexibility, and even the occasional failing.

'Sometimes I'll get the timing wrong, or something else in life will come up and I forget to add what's needed, or stir at the right time, bottle something at the time it needs.'

But that's par for the course in a busy home kitchen that Brydie describes as a constantly evolving beast. And the beauty of fermenting projects gone wrong is that you can just start over.

'I find it hard to stick to rules and precision, so flexibility is important. Some things work out, and others not so much, but at least I had a go. Good food and creativity are fundamentals to living for me, and being willing to fail is part of the creative process.'

> "I spent a lot of time in the kitchen as a kid, either stirring, rolling or simply watching my mother make everything from scratch ... I learned by osmosis really, simply being there and helping ...

CARING FOR YOUR FIRE CIDER

This stuff is not only germ proof, it's bulletproof. It'll last forever, and it's hard to mess up. Only the most basic rules of food care apply – don't store it in direct sunlight, and don't leave the lid off.

HOW TO

MAKE FIRE CIDER

CORE INGREDIENTS

- 2 large onions
- 200 g (7 oz) turmeric root
- 200 g (7 oz) garlic
- 200 g (7 oz) ginger
- 6 hot chillies
- Up to 1 litre (2 pt) raw unfiltered organic apple cider vinegar, with mother too (the protein, enzyme and bacteria-rich strands you'll see floating inside the bottle)
- 1 strip of lemon rind

OPTIONAL INGREDIENTS

- Horseradish
- Herbs such as rosemary and thyme
- Raw honey

TOOLS

- 1.5 litre (3 pt) watertight glass jar
- Chopping board
- Chef's knife
- Strainer
- Bowl

METHOD

1. Peel the onions, ginger, turmeric and garlic. Then chop, slice or dice all the whole ingredients into pieces roughly the same size.
2. Combine the chopped ingredients and any herbs you're adding in the glass jar and top it up with as much apple cider vinegar as it takes to cover the ingredients.
3. Store it out of direct sunlight and allow it to steep for 2–4 weeks.
4. Agitate the mixture by giving it a gentle swish around at least once every 24 hours.
5. After the fermentation period (minimum two weeks) has passed, strain the mixture into a bowl then stir through a generous spoonful or two of raw honey.
6. Tip the liquid back into the jar, or into a pour-friendly glass bottle, then store it in a dark corner.

TAKING YOUR TONIC

Some people take fire cider as a preventative during cold and flu season. Others are hooked on a fortifying daily dose – it's really up to you. Fermentation experts all agree on one thing: fire cider is just the thing when you feel a cold coming on. Take 1–2 tablespoons when symptoms first arise, and repeat every 3–4 hours until they subside. Don't bother diluting your fire cider with water; this beast is best taken neat.

USING THE LEFTOVER INGREDIENTS

Nothing needs to go to waste. The leftover chunks of steeped solid ingredients can be used to flavour your cooking – Brydie recommends stirring them through some dhal or a rice dish. You might also like to try using the tonic itself, perhaps as a base for salad dressing, drizzled on steamed vegetables and rice, in marinades, or even splashed into a Bloody Mary.

Step 1

Step 2

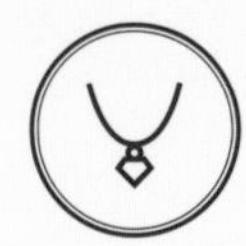

DIVE HELMET PENDANT

Jewellery manufacturing

Diving suits were invented in the early 18th century, to aid wreck diving and treasure hunting. Diving helmets came soon after; heavy contraptions made from copper or bronze that were fitted to full-length, watertight canvas diving suits and contained a valve that was attached by a hose to an air supply pump on the surface. This little treasure is a modern-day nod to dive helmets of old. It's a pendant designed to be worn on a chain around the neck (rather than carried on it), and is made from sterling silver by a jeweller based in Melbourne, Australia. It is crafted using the lost-wax metal casting process, where molten metal is poured into a mould that has been created around a wax model. Once the mould is poured, the wax model melts and drains away. But the best thing about this pendant is that the faceplate opens and shuts, just like it would have done on a real dive helmet.

MEASUREMENTS	Length: 2.7 cm (1 in); width: 2 cm (½ in); weight: 10 g (½ oz)
MATERIALS	Sterling silver
KEY TOOLS	Emery sticks, flexi drive with interchangeable hand piece, hand files, riveting hammer
KEY MACHINES	Centrifugal casting table, kiln, polishing motor, wax injector
TIME TO MAKE	10 hours
LIFESPAN	Heirloom quality

ORION JOEL — Manufacturing jeweller [Melbourne, Australia]

When Orion was a child living in far north Queensland in Australia, his father owned a 100-year-old pearl lugger, the *Floria*. Pearling was a major industry in the tropical regions of Australia from around the mid-19th to mid-20th centuries, and the *Floria* was one of its workhorses, carrying pearls back to shore after its crew of sailors and divers had raised them.

By the time Orion's father was captaining the boat it was a private vessel, and was filled with historical memorabilia. Orion spent time on the boat on weekends, and was fascinated by the old equipment and photos on display.

One item in particular, an antique diver's helmet made of copper and bronze that weighed around 30 kg (66 lb), captured his imagination.

Like all good ghosts, the helmet's memory stayed with Orion through childhood and adolescence, and into adulthood. When he set about creating a range of nautically themed men's jewellery, the dive helmet loomed large in his imagination.

He began by sketching out his ideas on paper, then used CAD software to create a three-dimensional master pattern. Then he made the mould master.

'With many of my pieces, I fabricate directly in metal or I hand carve the wax, but for the dive helmet, I made the master by printing a meltable wax and resin mould in 3D on a printer that's similar to a bubble jet printer, used the lost wax process to make the master, and then moulded it in a rubber mould,' says Orion.

Some believe the lost-wax metal casting technique was developed in ancient Egypt. It involves creating a mould around a wax model, heating the wax so that it melts away, then pouring molten metal into the cavity that remains. The ancients used wax, clay and bronze for the model, mould and finished piece; Orion uses wax, plaster and sterling silver.

'I create my moulds from plaster, heat them in a kiln at 800°C (1472°F) and, once the wax has disappeared, inject molten metal into the space it leaves, using centrifugal force. The metal takes the exact form that the wax left behind.'

Next, Orion cleans the cast piece using hand files and abrasive discs and wheels. It's here that he uses his favourite tool, a coarse hand file that he calls 'The Beast'.

'It's usually the first tool I use to start any job. I use it to set the shape and lay the platform of the piece before refining and finishing.'

Orion files it with polishing compounds and polishing mops on a Dremel and lathe and rivets its moveable face plate into place. Lastly, he oxidises it.

'I use a chemical – liver of sulfur – to oxidise the dive helmet. A chemical reaction with the small amount of copper in sterling silver turns the metal black, it's great to watch. Then I buff the pendant lightly with polishing compounds and mops to give it an aged, vintage look.'

Orion learned his trade working with his father and older brother, both design and manufacturing jewellers.

'What started as a casual school holiday job turned into a passion that has lasted for nearly two decades. After working for my father for 13 years, in 2013 I decided to go out on my own and moved from far north Queensland to Melbourne.

'My studio is one of many in a Victorian-era red brick building in Kensington that was a wool processing factory in the 19th century. It's small, just 4 sq m (13 sq ft), and I share it with another jeweller. It's filled with plants to keep the air fresh, and it has a rolling timber door that opens onto a cobblestone laneway. I'm here almost every day of the week, making beautiful treasures. I love it.'

> “What started as a casual school holiday job turned into a passion that has lasted for nearly two decades.

CARING FOR YOUR DIVE HELMET PENDANT

Take the pendant off for showering and sleeping. Avoid contact with chlorine and other chemicals; while this won't affect its durability, it may affect the finish.

To clean the pendant, brush it gently with an old toothbrush dipped in warm, soapy water. Then dry it thoroughly with a soft towel. Orion says this will bring it back up a treat, and is a great way to keep all of your jewellery clean.

Have a jeweller give it a check-up every three to five years to make sure the chain isn't wearing through the pendant bale (the loop that the chain goes through) and to check the locking mechanism of the moveable face plate to make sure it still clips in and out tightly.

HOW TO

MAKE A LOST-WAX CASTING

MATERIALS

- Wax – personal preference will come into play here. Some people like to use hard wax and carve it, while others like a very soft wax they can shape with their fingers. Orion uses custom jewellery wax, which can be bought online.
- Casting investment plaster
- Sterling silver casting grain
- Bucket of water

TOOLS

- Paper and pen or pencil
- A candle or alcohol lamp
- Wax carving tools or dental tools, files, sawblades, scalpel
- Heavy gauge metal flask with a rubber seal at one end – Orion uses old exhaust pipes that have been cut down and had a rubber cap attached, which can be bought online
- Old baking tray or cake tin
- Crucible (melting dish)
- Oxyacetylene torch
- Side cutters or small bolt cutters
- Hand file

MACHINES

- Kiln – preferably electric, though a wood-fired kiln will work too
- Spring-driven centrifuge casting machine
- Tumbler
- Flexi drive or Dremel

SAFETY GEAR

- Welding gloves
- Welding goggles
- Leather apron (optional)

METHOD

1. Sketch the model you plan to make on paper with a pen or pencil. It can be as simple or as complex as you like – a three-dimensional model of the entire *Titanic*, or just its anchor.
2. Light your candle or alcohol lamp and, taking care not to burn yourself, hold the wax above the flame and move it around until it is soft enough to shape with your fingers. If using wax wires or wax blocks, heating probably won't be necessary.
3. Now start to press and mould the wax into the shape of your design.
4. If you're attaching pieces of wax, repeat steps 2 and 3 and use the candle to soften the pieces where you want to join them together, then use your fingers to meld the wax one piece into another so they will hold together.
5. To further refine and shape your mould, use the wax carving tools to add texture and detail. Once you're happy with your wax model, attach it to the rubber base of the metal flask by spruing it – melting a small sprue (piece of wax) and attaching it to the model, then to the base. Place the flask over the top and insert it into the rubber base to seal the flask. One end will remain open.
6. Now prepare the casting investment plaster according to the manufacturer's instructions, and pour in enough to completely cover the wax model.
7. While the plaster is hardening, prepare your kiln by heating it to 800°C (1472°F) and placing an old baking tray or cake tin at the bottom.
8. When the plaster has fully hardened, remove the cap on the flask and place the flask, open side down, into the kiln. Take care to ensure the flask is positioned over the baking tray so that when the wax melts it is collected in the tray. When all the wax has gone, a negative image of the model will remain in the plaster.
9. Now place the flask on one end of the arm of a spring-driven centrifuge casting machine.
10. Put a crucible filled with the sterling silver casting grain on the other end of the casting machine arm.
11. Wearing your safety gear, use the oxyacetylene torch to melt the sterling silver, then loosen the catch on the spring-driven centrifuge arm to force the metal into the cavity.
12. Allow approximately 2 minutes for the metal to cool and harden, then take the flask and put it in a bucket of water to dissolve the plaster. You'll now have a rough sterling silver casting of your wax model.

HOW TO

MAKE A LOST-WAX CASTING

13. Remove the sprue from the base of the casting with side cutters or small bolt cutters, then use the hand file to smooth away any excess.
14. Now, place it in the tumbler for approximately 1 hour, which will pummel a nice polish into the surface of your piece.
15. Move on to polishing your piece with the flexi drive or Dremel, using successively finer abrasive wheels, then polishing compounds, to refine and finish the piece until you've achieved your desired finish.

Step 5

ANSFRID JUMPER

Knitting

When Vikings roamed the seas, they might have worn knitted jumpers (if you're speaking British English) or sweaters (if you're American) like these. And their descendants, and cold-climed people of many other places besides, still do. Knitted completely by hand in the fisherman's or Aran style (named for the Aran Islands on the west coast of Ireland) by a craftswoman living in the Loire Valley in France, this jumper is a fetching, 50/50 luxury yarn blend of Merino wool and Tibetan yak hair. Its style was inspired by the yarn itself, a high-quality fibre that is pleasant to work and ensures an excellent definition of the cable and Aran stitches it has been carefully crafted from. The button collar can be fastened with the metal, Celtic-engraved motifs buttons, or left open. Let's face it, if you can't catch a fish in this jumper, you probably never will.

MEASUREMENTS	Made to measure
MATERIALS	Buttons, luxury yarn (50 per cent merino wool and 50 per cent Tibetan yak hair)
KEY TOOLS	Cable needles, circular needles, knitting calculator, point protectors, scissors, straight needles, yarn needles
KEY MACHINES	None
TIME TO MAKE	2.5 days
LIFESPAN	Heirloom quality

SANDRINE FROUMENT-TALIERCIO — Craftswoman [Loire Valley, France]

Sandrine works in a small room in the house she shares with her husband and children in a village in the province of Touraine. Until the middle of the 19th century, the room – which measures just 10 sq m (108 sq ft) – was part of a sheepfold. It is superbly fitting that it is filled with wool today, albeit in the form of balls or skeins.

'I learned the basics of knitting with my mother when I was four and a half years old and more complicated techniques, like knitting in the round and cable knitting, with my grandmother a few years later,' says Sandrine.

She has knitted and crafted her entire life, but it wasn't until she had her first child in 2005 that Sandrine set her academic career (she has a degree in international trade and a PhD in ancient history) aside in favour of the life of a professional craftswoman. She founded her label, Atelier Lune de Nacre (Pearlescent Moon Studio), soon after.

A new project begins with sketches, then swatching. Here Sandrine experiments with the yarn and various needle sizes to obtain the best look and feel for the jumper, and to calculate the gauge (how many stitches and rows to the inch). Next, she uses a pencil, knitting graph paper, a tape measure and a knitting calculator to plan the pattern of the jumper in several sizes. Then it's time to start knitting.

'I start with the back and the bottom of the sweater, creating a ribbed edge with n° 4½ (US 7) straight needles a smaller diameter than for the body, so that the ribs are elastic. I continue with larger n° 5½ (US 9/UK 5) cable needles for the body of the jumper. Once the back is completed, I knit the front and the sleeves in the same way.'

Sandrine likes to keep the ball of yarn in a wooden bowl as she works, to prevent the ball from rolling away. She's also a stickler for point protectors – small, thimble-like domes that she puts over the ends of her needles when she sets her work aside. They protect the ends, and keep the stitches from accidentally slipping away.

> Through the window, I can see part of our field behind our house ... the meadow where our sheep graze, and two magnificent old trees ... I rise early to work, and every morning I see the sun rising over this landscape.

CARING FOR YOUR NATURAL FIBRE JUMPER

Wash the jumper by hand in lukewarm water with a mild detergent. Make sure the washing and rinsing waters are around the same temperature, to avoid felting. Don't soak the jumper for more than 10 minutes, and don't rub it during washing. To remove the water, squeeze the jumper without twisting the wool. You can also put the jumper in a large clothes net and put it in the drum of a washing machine on the wool cycle to wring it. Lay it flat to dry, away from excess heat and light. If pills (those pesky little balls of wool) appear, you can shave them off with a manual or battery-operated wool razor.

If you don't plan to wear the jumper for a long period of time, protect it against moths by placing it in a large zip-lock freezer bag and putting it in the freezer for 48 hours to neutralise any moth eggs, and then storing it – still in the bag – in a closet.

Once all the pieces are knitted, they are steam blocked to give them their dimensions and definitive appearance. Sandrine does this by gently pinning dampened cotton muslin cloth to the fabric with rust-proofed pins, then ironing it with a steam iron.

'Now the pieces must be sewn together. I pin them together with thin bamboo pins, and sew with a yarn needle. I begin with the shoulders, to join the back and the front, then the sleeves are joined to the sides. I continue with the bottom of the sleeves and the sides of the jumper.'

Next, Sandrine knits the button bands by picking up and knitting stitches along the neckline, then knitting two button bands and buttonholes. She sews the buttons on, then uses circular needles to knit the collar.

Lastly, she weaves in all the ends – again with a yarn needle – taking care to hide and secure the tail ends of the yarn.

Sandrine listens to cultural programs and music on the radio as she works, and drinks a lot of tea. When her eyes tire of her crafting, she looks out the window of her room.

'Through the window, I can see part of our field behind our house. In the background, our large permaculture-led vegetable garden and our fruit trees. In the foreground, the meadow where our sheep graze, and two magnificent old trees: a robin tree which embalms the garden in spring, and an immense spruce with a red grape vine climbing the trunk. I rise early to work, and every morning I see the sun rising over this landscape. It's always a wonder.'

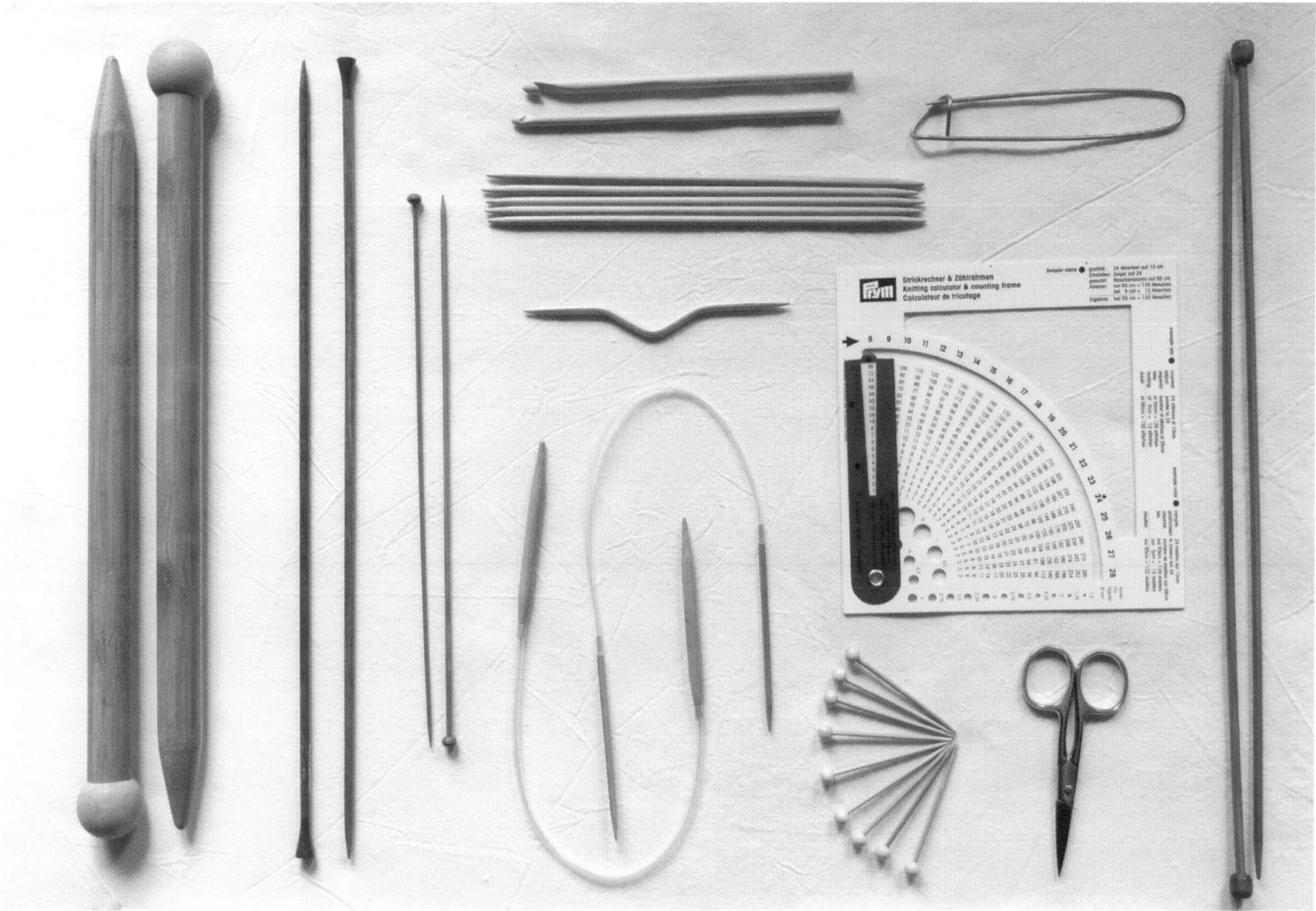
Prym
Strickrechner & Zählrahmen
Knitting calculator & counting frame
Calculateur de tricotage
Beispiel siehe
gezählt: 24 Maschen auf 10 cm
Einstellen: Zeiger auf 24
gesucht: Maschenanzahl auf 55 cm
Ablesen: bei 50 cm = 120 Maschen
bei 5 cm = 12 Maschen
Ergebnis: bei 55 cm = 132 Maschen

HOW TO

CAST ON

Sandrine shows us how to cast on using the thumb casting on method, and how to knit using a staghorn cable stitch. This is a wide pattern, worked over 16 stitches against a reverse stockinette stitch background, that is dramatic yet easy to produce. You'll need some knitting experience for the staghorn.

MATERIALS

- Yarn

TOOLS

- Straight needle
- Scissors

METHOD

1. Measure off a length of yarn about three times the length of the edge to be cast, cut it with scissors, then begin with a slip knot on the needle. Hold the needle in your right hand.
2. Tensioning the other end in your left hand, make a loop by wrapping the yarn around your left thumb. Bring the tip of the needle up under the yarn, alongside your thumb, and insert the needle into the loop.
3. Take the yarn around the needle, then slip your thumb out of the loop.
4. Tighten the loop on the needle so it is snug, but not tight.
5. Repeat steps 2–4 for each stitch, to the length of the edge of the piece you're creating.

Step 2

Step 3

Step 4

HOW TO

KNIT STAGHORN CABLE STITCH

MATERIALS

- Yarn

TOOLS

- Straight needle
- Cable needle

METHOD

1. Cast on 20 stitches (16 for the staghorn panel and 2 for a reverse stockinette stitch on each side of the panel).
2. First row: purl 2 – knit 4 – cable 4 back (slip next 2 stitches onto cable needle and hold at back of work, knit next 2 stitches from left-hand needle, then knit stitches from cable needle) – knit 4 – cable 4 front (slip next 2 stitches onto cable needle and hold at front of work, knit next 2 stitches from left-hand needle, then knit stitches from cable) – knit 4 – purl 2.
3. Second row: knit 2 – purl 16 – knit 2.
4. Third row: purl 2 – knit 2 – cable 4 back (slip next 2 stitches onto cable needle and hold at back of work, knit next 2 stitches from left-hand needle, then knit stitches from cable) – knit 4 – cable 4 front (slip next 2 stitches onto cable needle and hold at front of work, knit next 2 stitches from left-hand needle, then knit stitches from cable) – knit 2 – purl 2.
5. Fourth row: knit 2 – purl 16 – knit 2.
6. Fifth row: purl 2 – cable 4 back (slip next 2 stitches onto cable needle and hold at back of work, knit next 2 stitches from left-hand needle, then knit stitches from cable) – knit 8 – cable 4 front (slip next 2 stitches onto cable needle and hold at front of work, knit next 2 stitches from left-hand needle, then knit stitches from cable) – purl 2.
7. Sixth row: knit 2 – purl 16 – knit 2.
8. Repeat these six rows.

IN KNITTING CODE, THAT'S:

1. Cast on 20 stitches (16 for the staghorn panel and 2 for reverse stockinette stitch on each side of the panel).
2. First row: P2, K4, C4B, K4, C4F, K4, P2.
3. Second row: K2, P16, K2.
4. Third row: P2, K2, C4B, K4, C4F, K2, P2.
5. Fourth row: K2, P16, P2.
6. Fifth row: P2, C4B, K8, C4F, P2.
7. Sixth row: K2, P16, K2.
8. Repeat these six rows.

THE DERBY LAMP

Lighting

Let there be light, brother, o let there be light. People all around the world have been singing the praises of the sun, stars, fire and other illuminative forms of natural phenomena since, well, forever – and electric light too, ever since Thomas Edison had his lightbulb moment and invented the thing in 1879. This salvage lamp, the Derby, comes from Israel's design and art capital, Tel Aviv. It was made by a lighting designer and maker whose love of vintage items and commitment to sustainability in design led him to become a committed recycler and upcycler of things. It can be wall or ceiling mounted, or simply sit upon a surface, and is handcrafted from vintage Cuban cigar boxes, children's shoe lasts and a skateboard – all vintage items sourced in flea markets then meticulously restored and born again as lighting components.

MEASUREMENTS	Height: 80 cm (31 ½ in); width: 25 cm (10 in)
MATERIALS	Brass rods and joints, Cuban cigar boxes, LED lighting components, children's shoe lasts, vintage skateboard
KEY TOOLS	Crimper, cutter, dental tweezers, hammer, lighter, multi tool, pliers
KEY MACHINES	Dremel, drill, impact driver drill, sanding machine, soldering iron
TIME TO MAKE	10 hours
LIFESPAN	50 years +

RE'EM EYAL — Lighting designer and maker [Tel Aviv, Israel]

Most children are taught to steer clear of electricity, but Re'em Eyal made his first lighting fixture when he was just eight years old, after purchasing some electrical components from a dealer in south Tel Aviv. He'll readily admit it wasn't his most professional creation, but it sparked an interest that has lasted a lifetime.

Later, as an adult, Re'em qualified as an architect, but it was the structure of lighting, not buildings, that he decided to pursue professionally. He founded his business, Studio Oryx, in 2009.

The first stage in Re'em's lamp creation process is sourcing materials. The technical components for his lamps (wires, cables, switches and so on) are all brand new and of the highest quality, but he chooses to only work with second-hand or vintage items for the structure and aesthetics. This means he's always on the hunt for inspiring materials – in his studio you'll find an impressive collection of shoe lasts, cigar boxes, anything with wheels, lampshades and light stems, enamelware, woodworking planes, glass canisters and dispensers, and bicycle rims.

Re'em buys some things online, but prefers to gather his materials at one of Tel Aviv's renowned flea markets. He likes to examine things in person, for a start, zipping around the city on his 1962 Vespa or 1980 Honda motorcycle (two of his four collector's bikes that he has restored himself), hunting for treasure.

He takes two approaches to the design of his one-off lamps. He will either work with materials in his collection, or he will have an end design in mind, and search for the materials he needs to bring it to life. The Derby was one of the latter.

'I found the skateboard online. It's an American skateboard, dating from the sixties or seventies. It can take a month to find a pair of children's shoe lasts – these ones, I found in a flea market in Barcelona six years ago. I'll buy Cuban cigar boxes anytime I see them; I can always find a use for those.'

Once he has acquired all the materials, Re'em sets about preparing them for upcycling. He cleans any rust or patina off the brass with vinegar, polishes it, then applies a clear metal varnish with a brush and rag to revive and preserve the wood, metal and brass components. Where he needs to, he die casts, bends or welds brass components too, getting handy with a blowtorch, pipe bending tool and notching machine.

Next, he cleans and sands the cigar boxes and shoe lasts using a Dremel and sanding machine, then applies a clear wood varnish. He's now ready to start assembling the lighting fixture.

'It's only now that I'm starting to handle the electricity of the fixture. For this, I use pliers, a soldering iron, a crimper, a cutter, a hammer, a lighter, dental tweezers, a manual and an electric screwdriver, and a drill to safely and securely place the wires inside the lamp.'

Re'em really enjoys the delicate work the electrics demand. 'I especially like using the dental tweezers. Working with the tweezers makes me feel like a surgeon trying to rescue his patient.'

Finally, it's time to put in an LED bulb and test the light. Re'em does all of this work in a 100 sq m (1076 sq ft) studio that was once a carpentry workshop, located in Tel Aviv's Florentine neighbourhood; once Tel Aviv's centre of industry and now one of its main creative hubs.

'I love the whole process of creating a light. From my interactions in the flea markets to the process of reviving the elements of what I find and reassembling them into something new and old at the same time – almost everything can be repurposed and used again, if you have the motivation.'

"

I especially like using the dental tweezers. Working with the tweezers makes me feel like a surgeon trying to rescue his patient.

CARING FOR YOUR SALVAGE LAMP

Change the lightbulbs as needed. Re'em recommends using LEDs, because of their low energy consumption. Keep the lamp clean with a feather brush and wet wipe towels.

HOW TO

REMOVE RUST AND PATINA FROM METAL

MATERIALS

- Metal to be restored
- White household vinegar
- Polishing soap
- Metal varnish

TOOLS

- Spray bottle
- Brush or cloth

MACHINES

- Brushing machine or steel wool
- Polishing machine

METHOD

1. Fill a spray bottle with white household vinegar.
2. Spray some on the rusted area and let it sit for a few minutes.
3. Place a soft brushing wheel on the brushing machine then switch it on. As the brush turns, hold the metal against the brush and turn it until the rust has been removed. If you don't have a brushing machine, use steel wool.
4. To remove patina, use either steel wool or a polishing machine to achieve your desired finish. The polishing machine works in the same way as the brushing machine; it will result in a matt finish. For a glossier finish, simply hand scrub the metal with steel wool.
5. To bring out the metal's shine, apply the polishing soap to the polishing machine's wheel so that it has a good covering, then polish the brass on the wheel as in step 4.
6. To retain the metal in this condition, use a brush or cloth to wipe on a layer of metal varnish.
7. If you'd prefer the metal to oxidise and develop a new patina over time, don't bother with the varnish.

Step 2

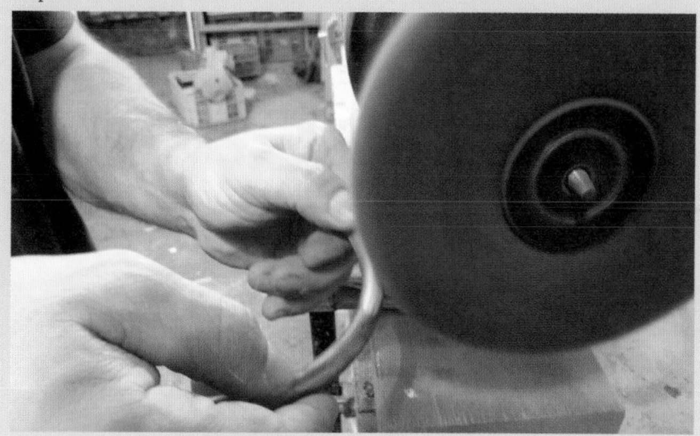

Step 3

Habanos s.a.
HECHO EN CUBA
Totalmente a mano

PERSEVERE LETTERPRESS PRINT

Printing

Some things are better the old-fashioned way – and one of those things is printing. For nigh on five centuries, letterpress printing, a relief printing technique that works by pressing an inked, raised surface against paper, was the dominant form of printed ephemera. Invented by Johannes Gutenberg in Germany in the mid-15th century, the press enabled the mass production and dissemination of printed materials, and is considered one of the most significant inventions of the second millennium CE. Letterpress printing was usurped by offset printing in the 20th century – well, almost. A resurgence in small-scale, artisanal printing has kept the craft alive. Jennifer Farrell of Starshaped Press in Chicago is one such revivalist. This print, Persevere, dates from 2016, and could be an ode to letterpress printing itself. The word 'persevere' is set entirely in tiny metal type and ornaments to create all of the letterforms, some of which are over 100 years old.

MEASUREMENTS	17.8 × 12.7 cm (7 × 5 in)
MATERIALS	Carbon paper, metal type, wood type (background) ink, handmade paper by Porridge Papers (US)
KEY TOOLS	Composing stick, galley, wood and metal type
KEY MACHINES	10 × 15 Chandler & Price press, guillotine
TIME TO MAKE	4 hours
LIFESPAN	Heirloom quality

JENNIFER FARRELL — Printer [Chicago, USA]

Jennifer Farrell opened her studio, Starshaped, in 1999, and likes to say that she prints like it's 1929. It seems to be true. You won't find any computer-set type or design shortcuts here. She sketches her ideas on paper and works with type that is sometimes more than 100 years old.

'I begin with an idea and sketch it out loosely on graph paper to figure out which type and ornament will best fit the desired size. Then I look through my type specimen book and the type cases themselves to check for size and characters to see how it will fit within my idea,' says Jen.

Then she pulls out all of the type she needs and sets it in a composing stick, along with the ornaments.

Sometimes necessity – such as some old type missing characters, or being unable to change the point size of metal type – demands that the final design veers from the concept. Jen is open to the alternate outcomes and happy accidents this can result in.

'My favourite part of the process is the actual typesetting. I love the Tetris-like skill needed to get all of the pieces together and squared up in order to go on the press.'

'I place all of the type on a galley (a metal tray) and pull a galley proof on the proof press with carbon paper. If I'm not satisfied I make changes until it's perfect. Then it is locked up in the press to be printed with the colour of my choice on the paper I want to use.'

Last time she counted them, Jen had more than 1000 metal and wood typefaces in her studio.

'They come from a variety of sources: there are five type foundries in the US that still sell type that I get many items from; some are gifts, and some are serendipitous finds in antique stores.

'I cut the paper on a guillotine ahead of time to the approximate size I need – sometimes the exact size, depending on the print – and determine how many copies I'd like to print. I allow for extras in

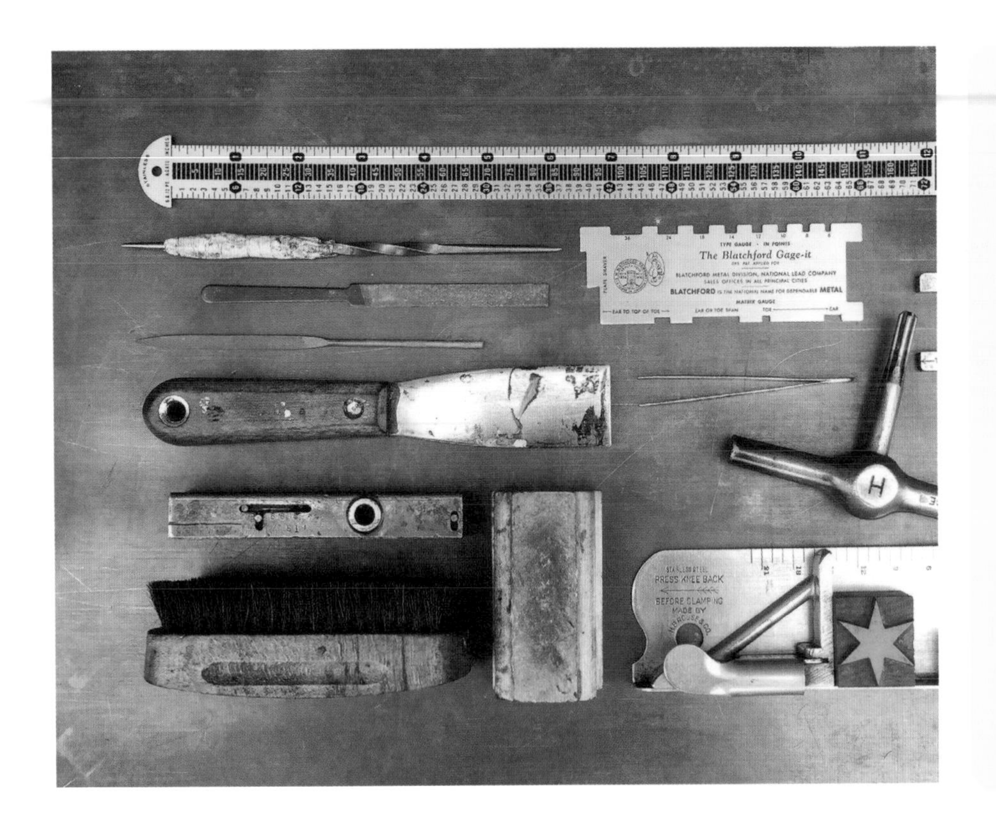

CARING FOR YOUR LETTERPRESS PRINT

Prints are not made to last forever, but some simple principles of care will greatly increase its lifespan. Keep your print in a temperature-controlled environment that is neither too extreme nor too humid. It should be stored flat, ideally with archival tissue between it and other prints. If the print is to be framed, it's best to have this done professionally by someone using spacers (to keep the print from touching the glass and archival mats), and conservation glass (to help cut down damage from light exposure).

the set-up process, as some are lost while adjusting the press for the final run.'

Once the set-up is ready to go, Jen inserts each print into the press one at a time, until the edition is finished.

Her 74.3 sq m (800 sq ft) studio is one of around twelve small businesses and artisans in a two-storey deco building in the Ravenswood neighbourhood of Chicago.

'In the last ten years many restaurants have opened and that means people from all of these spaces are out walking around and seeing each other every day. That's pretty nice.'

Starshaped takes on all sorts of custom work, from business cards to music packaging, social stationery to posters, limited edition prints, books and greeting cards.

'I design all of our work, and it's all done with materials in the studio; metal and wood type, found images and linoleum cuts.'

Jen likes to work in the evening, listening to music while she works.

She believes that attention to detail, an organised approach to a workspace and the ability to maintain design principles while working with a hands-on medium are the best skills for a letterpress printer to have.

Jen first got into printing when she was in college, when there was a big push to develop websites and move all design onto a screen. What attracted her to the craft then is what keeps her going now; it provides a tactile outlet for the work of her hands and a set of challenges that push her design work forward.

'Instead of pushing type around on a screen you're pushing it around right in front of you and connecting your own design work to that of printers of the last 50 to 150 years. The traditional methods combined with the new, as well as emerging ideas about design, mean that it's a viable craft.'

“

My favourite part of the process is the actual typesetting. I love the Tetris-like skill needed to get all of the pieces together and squared up in order to go on the press.

HOW TO

MAKE YOUR OWN LINOCUT RELIEF PRINT

In practice, letterpress also includes other forms of relief printing, such as wood engravings, photo-etched zinc cuts (plates) and linoleum blocks. This project uses a sheet of linoleum as the relief surface.

MATERIALS

- Tracing paper or opaque layout paper (waxed kitchen paper will work)
- Graphite transfer paper
- Linoleum block (you can buy lino on blocks, or glue a sheet of lino to a wooden block)
- Water-based ink of your choice
- Paper to print onto

TOOLS

- Pencil
- Lino cutter
- Chisel or gouge
- Brayer or roller
- Spoon

METHOD

1. Using the tracing paper, draw your text or image as you want it to appear on your final print.
2. Once your design is ready, lay a sheet of graphite transfer paper over a linoleum block and then lay your tracing paper on top, with the text or image face down. This will give you a mirrored template you can transfer onto the lino.
3. Fasten the carbon paper and the tracing paper so they can't move about, then use the pencil to trace over the lettering image, following the lines as closely as you can. This will transfer the design onto the lino.
4. Remove the layers of paper; you should now have a perfect template for your design.
5. Take the cutting tools and gradually cut away the areas that won't be printed. Begin by using the cutter to cut around the shape that you'll retain, then move on to removing the excess lino with the chisel or gouge.
6. Use a brayer or roller to ink the block.
7. Lay your paper down on the block, then use the back of a spoon to burnish (rub) where the image is and transfer it to the paper. Make sure you rub the whole image, but don't press too hard lest you mark or tear the paper.
8. Carefully remove the paper from the block – remember, the ink will be wet, so don't slide it but try to lift it off as cleanly as possible – then lay it aside to dry.

Step 3

Step 4

Step 6

VANS
GASOLINE
MOTOR CO.

GASOLINE
CO.
GAS
THIS SIDE UP
Kawasaki

XX MOTORCYCLE

Engineering

This custom motorcycle, the XX, is inspired by traditional American flat tracker motorcycles that tear up dirt, track and road in track racing, an old-school motorcycle racing series that dates back to the 1930s. The XX was built by a pair of gun motorcycle technicians at Gasoline Motor Co. in Sydney, Australia: 'the contemporary institute of recreational and mechanical innovation'. Gasoline can convert any motorcycle into something original and custom, and make it look damn fine in the process. The XX is named after the Roman numeral for 20, which is its racing number. It features a 1200cc Harley Davidson Sportster engine and a 27-part raw steel exhaust, which was handmade and welded piece by piece, from engine to tip, by the Gasoline crew themselves. It's made for a dirt flat track, but the XX is street legal.

MEASUREMENTS	Height: 1.4 m (4 ½ ft); length: 2.2 m (7 ft); weight: 200 kg (441 lb)
MATERIALS	Aluminium, carbon fibre, copper wire, fibreglass, forged steel, Harley Davidson XR 1200 Sportster (donor bike), leather, mild steel, stainless steel, paint, plastic, rubber
KEY TOOLS	Allen keys, drills, hand tools, hammers, screwdrivers, sockets, spanners
KEY MACHINES	Metal grinder, metal saw, TIG welder
SAFTEY GEAR	Fireproof gloves, goggles, mask (for grinding and welding)
TIME TO MAKE	15 days +
LIFESPAN	Heirloom quality

JASON LEPPA AND SEAN TAYLOR — Motorcycle technicians [Sydney, Australia]

Jason Leppa (head of creative) and Sean Taylor (master builder) have known each other since 2014. They met when Sean approached Jason for some Saturday work at Gasoline Motor Co., which Jason opened back in 1994. While working together, the pair soon realised that Sean's technical prowess and Gasoline's focus on motorcycle creativity and innovation were a perfect fit.

Gasoline builds around 45 motorcycles a year for their customers, and another five or so for their own enjoyment. They also service motorcycles and scooters, and sell helmets, apparel, and pre-loved vintage motorcycles of all makes and models. Their workshop is in Waterloo, in the heart of South Sydney.

'Once the area was an industrial estate but it's become a hub for growing creative businesses, like Gasoline. Our workshop consists of nine bays, where we could be knee-deep into nine builds at any one time. We have four very large toolboxes we go to for the tools we require for a particular job.'

There is one toolbox for each of Gasoline's technicians. Each holds all the hand tools they might need, from screwdrivers and spanners (wrenches) to socket sets and hammers: often in both imperial and metric.

'Every custom bike has its own processes and stages. We begin with a conceptual design phase where we come up with the shape, colours, style, and performance work – how we want the motorcycle to perform on the road and on the flat track,' explains Sean.

Next, Jason and Sean 'dummy build' the motorcycle with parts they already have or have sourced especially, and fabricate and manufacture parts that are missing to complete the look of the motorcycle.

Then they strip the motorcycle down to a complete shell and bare frame to start cleaning and painting parts.

'We work very methodically, moving step-by-step towards the way the bike will look at the end of the build. We ensure we have everything covered by ordering our parts; to work effectively, mechanics have to be very organised,' says Jason.

The pair fabricate a complete exhaust system by hand, using a TIG welder, a metal saw and a metal grinder.

'The XX's exhaust system is one of the most intricate we've ever done. Each piece has been individually cut to fit, and the welding was done using aircraft quality TIG welding techniques.'

Next, they finalise the design for the paint scheme and begin the painting process. They take the bike and the design to their painter, then work alongside him to tape the bike up with masking tape in preparation for painting.

The bike is returned to the Gasoline workshop, then it's time for the final assembly. Jason and Sean make sure the parts are going to fit and the motorcycle is going to work. They complete the internal wiring and fit the electrical componentry, test and tune the engine, and then it's time for road testing and final tweaks.

'We are big fans of American flat tracking. We'll both be out on the race track on the XX when the machine is complete; we always test our machines to full throttle ability, to make sure the engine's performance is up to standard.

'People often forget to stand back and check the lines when they're working on a build. They'll throw a tank on, throw a tail on and finish it off and think it looks a bit weird; it's because they haven't stepped back to review their work. Sometimes we'll stand around the bike with our arms crossed for an hour, thinking about how we can make it better.'

CARING FOR YOUR CUSTOM MOTORCYCLE

Ride and clean and have the motorcycle serviced at its appropriate intervals. For a bike like the XX, this means riding it at least fortnightly, keeping the engine and rims clean (if you're riding on a dirt track, do this after every ride), and having it serviced every 2000 km (1250 mi).

"

Sometimes we'll stand around the bike with our arms crossed for an hour, thinking about how we can make it better.

HOW TO

DETAIL YOUR ENGINE

A motorcycle's engine should be thoroughly detailed (degreased and polished) once a year. This will take a couple of hours.

MATERIALS

- Motorcycle to be cleaned
- Water-resistant tape such as gaffer or electrical tape
- Degreaser
- Cream-based metal polish

TOOLS

- High-pressure hose or garden hose
- High-pressure air gun or air compressor with hose attachment (optional)
- Clean rags – a few old T-shirts will do

METHOD

1. Find an area that allows for water and grease run-off and complies with water regulations in your area. A car wash bay is a good bet.
2. Park your bike. Seal off any open areas on the bike, namely the carburettor and air filters, and the switch block for the electrics on the handlebar, using the water-resistant tape.
3. Spray the engine with can of degreaser – use the whole can if you need to – and leave it for 10 minutes to settle in.
4. Now use a high-pressure hose (if you don't have one, placing your finger over the end of a garden hose and turning the water pressure up will work, at a pinch) and hose off the dirt and grease. Avoid spraying the hose directly on the areas sealed by the water-resistant tape.
5. If you have a high-pressure air gun or air compressor with a hose attachment, use it to dry the motorcycle, particularly the areas you'll struggle to reach by hand, like the brakes and undercarriage.
6. Now take the cream-based metal polish and apply it thoroughly to the motorcycle's metal casings. Jason prefers to apply the polish with his fingers, rather than with a cloth, as it helps him get the polish into all the nooks and crannies. The polish will turn black as you rub it in – this means it's lifting off the tarnish.
7. Once all the casings have been covered, take the clean rags and start polishing the metal. Get your back into it; this is the most intensive part of the detailing.
8. Wipe the entire motorcycle down.

Step 7

UNION

UNION NEON SIGN

Sign making

Neon signs were developed in France in the early 20th century. You know the ones – electric signs lit by long, luminous gas discharge tubes that contain rarefied neon or other gases. Those bright, unblinking signs of day and night that advertise wares, or locations, or thoughts and feelings, with retro-tinged aplomb. You no doubt have one or two heritage-listed numbers in your hometown; if not, just look to Las Vegas. This one was designed by a self-taught sign designer and fabricator near Ottawa in Ontario, Canada, who is inspired by a time when signs were unique, well designed, and built to last. That time was the 1920s–1960s, but rest assured, even if neon signs aren't as commonplace as they used to be, their luminescence, longevity and downright stylishness mean nobody's calling lights out on neon just yet.

MEASUREMENTS	1.8 × 0.8 m (6 × 2½ ft)
MATERIALS	Glass, steel
KEY TOOLS	Beverly shear, pencil, stomp shear
KEY MACHINES	Nil
TIME TO MAKE	100 hours
LIFESPAN	50 years +

SCOTT ADAMSON – Specialty sign designer and fabricator [Ottawa, Canada]

Scott Adamson turned to sign making in 2013, after his years as a deskbound graphic designer and photographer started to wear thin. He spent most of 2014 devouring books on the topic and practising; not long after, he purchased an entire shop's worth of equipment from a retiring neon bender. For Scott, making neon is 80 per cent mad science, 20 per cent magic.

'The process of making neon is very intimidating. You're working with fragile leaded glass tubing, molten glass, torches and burners, a variety of gases, and extremely high voltage electricity,' says Scott.

Scott has always loved old signs. He had a hunch there was a niche in the sign business for someone who built well-designed, high-quality signage using old techniques in a modern way. He was right. Since 2013, when he launched his business, Gaslight, he's gone from working solo in a downtown apartment to working out of a commercial space on the outskirts of Ottawa with a project and sales manager, fabricator, painter/finisher, and an install team.

Every project Scott works on is unique, and has its own conceptual design phase. Here Scott sets the direction of the project, and creates general concepts that work with that direction.

'Design is the fun part; it's where all of my wildest dreams can come true. My pencil is the medium between my brain and the physical world. I can create anything I imagine; it has unlimited potential. The rest of the tools just serve to build whatever the pencil comes up with.'

'Once we've narrowed the design down, we need to do some engineering; work out how to actually create the sign. City regulations and bylaws can really kill the fun here, but they have their reasons, and after a while you learn to build cool signs within the rules.'

Next comes the fabrication phase; building the sign. Scott likes to keep things low tech and manual wherever possible. The Union sign was made in 2015 for an advertising agency in Toronto.

'All of our dimensional signs are cut from a blank sheet of steel or aluminum. We quite literally craft raw materials into a finished product. With neon, we stock straight lengths of glass tubing. The colours are created based on how the type of powdered coating inside the tubes reacts with the types of gases that get pumped inside and ignited with high-voltage electricity.

'I regularly use a stomp shear that was built in 1955 and a beverly shear built probably not long after. These tools were built to last and I like to think they imbue that quality into the projects they're used on.'

Once the sign is built, it's sprayed or painted. The team used to do this themselves, but they've recently partnered with a professional paint shop in the area.

Last but not least comes delivery and installation. 'Every install we do presents its own challenge. We never encounter the same scenario twice; there's always a bit of detective work that comes into play. We generally want to secure the signs to a stud or structural element of the building. This is sometimes easier said than done. Often we are installing in a finished building so we need to find the best way to anchor the sign and run power to it. Over time we've learned it's best to sort all this out at the very beginning, during the design phase. It saves us a lot of headaches!

'The funny thing about signs is you never really appreciate them until they're gone. There are so many great signs that have been torn down over the years and replaced with something cheap and temporary. I'm sentimental by nature and I miss seeing those elaborate, awe-inspiring signs of yesterday. If I can bring back even a fraction of that feeling for someone someday, I'll feel like a success.'

CARING FOR YOUR NEON SIGN

Components of a neon sign have certain life spans and need regular refreshing. Incandescent bulbs will last six months to a year. LED bulbs will last four to five years and neon will last five to ten, depending on usage. The other main reasons a sign might stop working are the elements, and physical damage. To keep these wolves from the door, keep the sign sheltered from sun, wind, ice and rain and keep it in a place where it can't easily be physically damaged.

"

Design is the fun part; it's where all of my wildest dreams can come true. My pencil is the medium between my brain and the physical world.

HOW TO

MAKE A NEON LOOKALIKE SIGN

Sorry, champ. A do-it-yourself neon sign project is just not possible, or safe, for most of us. But this crafty project might be the next best thing – a neon lookalike that you can knock up yourself.

MATERIALS

- EL wire
- 16-gauge metal wire

TOOLS

- Large piece of paper – it's fine to use scrap, as this won't be part of the final sign (we tacked together two pieces of A3 art paper)
- Ruler (optional)
- Pencil
- String or wool
- Measuring tape
- Pliers or wire cutters
- Hot glue gun
- Sticky tape, e.g. washi tape
- Clear tacks (optional)
- Velcro or Command strips (optional)

METHOD

1. Draw the words or shape you plan to illuminate onto a large piece of paper. Write in cursive script – the letters and words need to be connected. If you are a stickler for straight lines, mark out a top and bottom border using a pencil and ruler first, to guide your letters.
2. Use the string to trace around the letters, then remove it, straighten it out and measure it to work out how much wire you'll need.
3. Now take a length of 16 gauge wire the same length as the piece of string, and a pair of pliers, and start bending the wire into the shape of your letters. This will form the frame of your sign. Remember, you'll need to use one continuous piece of wire for the entire phrase.
4. Once you're finished shaping the wire, tie down any loops with a small piece of wire to help the wire lay flat. The wire can be sharp, so take care to bend the ends down so they won't scrape the wall when you hang it up. Your wire frame might pop up in places. If this happens, make small bends to make the sign lay as flat as you can – it doesn't have to be perfect, but you will need a semi-even surface to place the EL wire on.
5. Take your EL wire and hot glue gun, making sure the wire's battery pack is positioned off to the side and will work in the place you plan to hang your sign. Begin gluing the EL wire onto the wire frame, bit by bit. Once you've applied a line of glue, hold the EL wire to the metal wire for 5 to 10 seconds, to be sure it stays.
6. Leave the glue to dry, then hang the sign on your wall – or wherever you plan to hang it – with some sticky tape or clear tacks. If there's a surface within reach, the battery pack can rest on that, otherwise you can use Velcro or Command strips to attach it to the wall.

RED SOX
ELECTRIC

INDUSTRIAL
SEWING
WORKSHOP

THE COMMUTER PANNIER BAG

Industrial sewing

A pannier is a bag, basket, box or other container that can be slung, either individually or in pairs, over the back of a beast of burden – think camels, donkeys, horses – or attached to the sides of a bicycle or motorcycle. The term has its origins in a Middle English borrowing of the Old French panier, which means 'bread basket'. This bicycle pannier is made from 340 g (12 oz) polycotton coated canvas that has been waterproofed for all-weather cycling and wear. Its unique attachment system means the hooks stay on the pannier rack, not the pannier itself, making the bag comfortable for carrying off-bike, too. Other subtle but thoughtful design touches increase the pannier's durability, such as the two-layer canvas lining: the seams don't line up, so any water that does make it past the canvas can't travel directly through. The pannier also has replaceable parts, like the liner and bungy cord loops and straps. Genius.

MEASUREMENTS	Height: 41 cm (16 in) closed, 58 cm (23 in) open; depth: 17 cm (7 in) at base, 20 cm (8 in) midway; width: 35 cm (14 in) top, 30 cm (12 in) bottom; volume: 25 litres (7 gal); weight: 1.3 kg (3 lb)
MATERIALS	Aluminium (hanging system and bar), bonded nylon thread, buckles, polycotton (rear canvas, reflective tape), polycotton heavy duty 340 g (12 oz) canvas (bag body), polyester (binding, seatbelt webbing, shockcord, zip), stainless steel (bolts, eyelets, nuts, washers), woven polyester (label), 170 g (6 oz) proofed canvas (inner)
KEY TOOLS	Anvil, awl, bone folder, hand chuck and die, hammers, hole punches, jig, mallet, mini thread nippers, scissors, vice
KEY MACHINES	Cylinder arm with synchronised binder, electric rotary cutter, electric straight stitch sewing machine, heavy duty walking foot sewing machine, heat cutter, kick press, metal parts tumbler
TIME TO MAKE	2 hours
LIFESPAN	10 years +

CATHY PARRY — Industrial sewer [Castlemaine, Australia]

Canvas is sturdy stuff. This plain-woven heavy duty cotton fabric has long been used to make tents, sails, marquees and backpacks. Even when unproofed, canvas is naturally water resistant, as the fibres swell when wet. Industrial sewer Cathy Parry knows all this, which is why it's the chosen medium for her industrial-strength pannier bag.

There are more than 40 steps in making this pannier bag, and Cathy uses a lot of muscle power putting the canvas through the gauntlet of her industrial sewing machines.

She begins by marking and cutting the fabric with an electric rotary cutter, often preparing enough pieces to make several bags at a time. Then she uses a hot knife to cut and prepare the webbing straps and loops, and sews them on using the walking foot machine.

Next, Cathy prepares the back of the pannier on the walking foot, and inserts eyelets using the kick press.

Cathy's machines come from all over the world – Taiwan, East Germany, Great Britain, Switzerland, Japan, the USA, Germany and Australia.

'I love my machines and tools; I can't stop collecting them! Today my favourite is my kick press. It's a cast iron eyelet press that stands on the floor and is operated by kicking and pushing with your foot. I can insert eyelets, studs and rivets with it perfectly, and beautifully.'

After using the kick press, Cathy sews up the bag and liner, joins the two and turns it the right way out – no mean feat, given that the

> I love my machines and tools; I can't stop collecting them! Today my favourite is my kick press ... I can insert eyelets, studs and rivets with it perfectly, and beautifully.

canvas often grazes Cathy's knuckles as she pulls and coaxes the fabric into place.

Once the bag is completed, Cathy smooths the rough edges on the metal parts for the hanger by tumbling them, then hand bends them using the jig and vice. The set – the pannier and its rack – is then ready to send to its new owner.

A family history of making and a lifetime of crafting led Cathy to become an industrial sewer.

'I got a job in a small factory around the corner that made truck tarps, awnings and industrial bags. I went in one day to get offcuts of fabric for an artwork, then went back the same afternoon to get eyelets put into the pieces. The owner was so impressed he offered me a job. It was there I learnt about the qualities of canvas, and developed my obsession with offcuts,' says Cathy.

Today, Cathy works out of a 75 sq m (807 sq ft) warehouse space in an industrial estate in Castlemaine, a country town turned creative hub some 120 km (75 mi) north-west of Melbourne.

'The space has a mezzanine for fabric storage, a big roller door and no windows, so we – me and my partner Chris Guest, who is an architect and also designed the pannier – made a big clear plastic tarp attached to a frame for the doorway. This lets heaps of light in and keeps out the weather.'

As an industrial sewer, it helps to be able to think things 'inside out', as that's how things are usually constructed. A flexible attitude, tough fingers and body awareness also helps, says Cathy, who spends a lot of her time working out how to overcome obsolescence in her products.

'For the pannier, we deliberately built in replaceable parts like the bungy cord loops and straps. These can be replaced by the owner of the bag when that part wears over a number of years. The liner is replaceable too, and can be taken out to access parts of the bag that might need repairing. Someone could do this themselves if they had the right equipment, otherwise any industrial sewer could work it out.'

Her favourite part of the pannier making process?

'The very end when I finish sewing, add the straps and fold the flap over. I've finished all the work, the canvas looks amazing and the bag is about to go out and have its own adventures. These things are made to be used, and used hard. I love that.'

CARING FOR YOUR CANVAS PANNIER

With proper care and respect and the timely replacement of spare parts, a canvas pannier will last for many years. Bear in mind that it's made from fabric, just like clothing. Take care not to get it too dirty, and give it the occasional day spa by removing dry dirt with a soft brush, soaking the bag in warm water and lightly brushing dirty patches with a soft brush, then hanging it and allowing it to dry completely before using it. Never scrub it with a heavy scourer or brush, as you will damage the print. If the rain stops beading on the canvas, it's time to reproof. Purchase a protective spray from any good camping store and follow the instructions to keep your pannier shipshape.

HOW TO

MAKE A BICYCLE PANNIER

Cathy talks us through how to repurpose a simple canvas bag as a bicycle pannier. You'll need to have a pannier rack attached to your bike.

MATERIALS

- A simple canvas bag
- 2 old leather belts
- Thin piece of plywood, mdf or plastic sheet (no more than 3 mm/⅛ in)
- 2 old belt rivets
- Bolts and nuts
- Hook and loop tape or extra belt

TOOLS

- Knife for cutting leather (such as a snap-off blade)
- Awl
- Drill

METHOD

1. Measure the length for your belts. They will be attached to your pannier, and be used to attach the pannier to your bike. You can work out how long to make them by threading them through the hanging system and holding them roughly where they will attach to the pannier. This will show you how long they need to be.
2. Use the snap-off blade to cut your belts to length.
3. Cut the plywood (or other material) to fit inside the back of your bag. This will keep the back of the bag stiff, which stops it from fouling your wheel.
4. Working with the bag inside out, use the awl to push through the fabric at each of the four corners.
5. Now use rivets to attach the plywood to the bag through the holes you've made.
6. Next, attach the belts to the top of the back of the bag. Place them as wide apart as is practical (this will help balance the pannier when it's on the bike), then drill a hole through the belts, the back of the bag and the board.
7. Working from the outside in, push the bolt through the hole you just drilled and add a washer and nut on the inside of the bag. Add the hook and loop tape or extra belt at the bottom of the bag in the same way so the bag doesn't swing out at the bottom when you go around a corner.
8. Attach to the rack at the three points, and away you go.

PRO TIP

Make sure the back of the bag has no straps – you really don't want straps to get caught in your wheel. If your bag is on the large side, angle the bottom of the bag towards the back of the bike so your foot doesn't hit the bag when riding. Make sure you tuck any hanging straps up and out of the way, so they don't get stuck in the wheel.

40
和紙

奉拝
熊野本宮
平成二十八年元旦

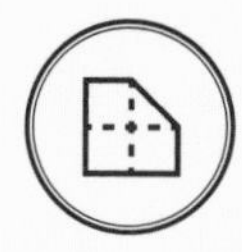

WASHI PAPER

Papermaking

Japanese kamisuki shokunin, *or paper makers, have been making washi paper entirely by hand using chemical-free methods for thousands of years. Made from natural wood pulp, washi's characteristically long fibres make the paper incredibly hard wearing, fire and water resistant, and practically impossible to tear. Exposure to sunlight is also no problem for this wonder paper; faced with blinding rays, washi simply whitens further. It's not surprising then that specimens of washi remain in Japan today that are more than 1000 years old. Washi's diverse uses include sliding doors, certificates, sculpture, painting, postcards, writing paper, origami and banknotes. This paper is made in a workshop on the main street of a town called Hongu, in Kumano, a historic sacred region and World Heritage site on the main island of Honshu in Japan.*

MEASUREMENTS	A4 (21 × 29.7 cm/8 × 12 in)
MATERIALS	Wood pulp, *neri* (a natural adhesive), ash
KEY TOOLS	*Kushi* (a large, rake-like tool), *masa* (tool for stirring the pulp), *sugeta* (frame for sifting paper), *sukibuné* (tub where washi is made), *takuri* (harvesting knife)
KEY MACHINES	*Dasuiki* (drying machine)
TIME TO MAKE	2 hours
LIFESPAN	10 years +

FUCHIKAMI-SAN — Washi paper maker [Hongu, Japan]

Washi paper making is an intensive business. It's equal parts preparation and creation, and it can really test a maker's patience, willpower, and muscle power. Luckily, Fuchikami-san, who came to his trade after years as a professional stonemason, has mastered the art of physical and mental concentration.

Washi is made in several steps: harvesting the plant, making the pulp, and finally making the paper. It all begins in the forest. Japanese washi paper is made from the bark of the Oriental Paperbush Mitsumata (*Edgeworthia chrysantha*) and Japanese Kouzou (*Broussonetia papyrifera*) trees, which grow in the nearby Kumano mountains.

Maintaining the trees brings Fuchikami-san great joy. He spends time in the fields, trimming trees at least once a week, to help them to grow straight and tall. He harvests the bark annually, in December and January.

He takes the bark back to his workshop, cuts sections of the trunk into equal lengths and steams them in an open-fire barrel for about four hours. Before it cools, the bark is stripped from the trunk, then soaked in water. This softens it and makes the next stage easier: using a knife to strip the dark outer skin from the bark.

The remaining white inner-layer, called *shirokawa* (white bark), is what is used to make washi paper. Fuchikami-san strings it outside in the sunshine and leaves it to dry for up to three days.

Next, he puts it in a large pot, adds ash and water to the dried white bark to form a lye solution, stirs it with a tool called a *masa*, and boils it for around three hours.

'This softens the bark, separating the fibres and removing any excess unwanted material. After this I soak the white bark in water from the Otonashi River, which runs behind my property, for around 24 hours to wash away the ash and any remaining dirt. This whitens the white bark fibres even more.'

Fuchikami-san carefully checks the white bark, removing any flawed or dirty pieces by hand, to ensure the paper will be as white as possible.

'Next, I beat the cleaned bark with a hard oak stick for two hours. If the bark is not struck firmly and thoroughly enough, good quality paper cannot be made. When it's ready I put it into the *sukibuné* (tub), along with water and *neri*, a natural glue made from the roots of the Tororo aoi (*Abelmoschus manihot*) plant and stir it with a *kushi* (a large rake-like tool).'

Finally, it's time to make the washi paper. Fuchikami-san does this by taking a sieve-like tool called a suketa and scooping up pulpy material from the vat. He moves the sieve back and forth to drain away excess water, then spreads the pulp out into a thin sheet.

'After the pulp has been sieved into sheets, I roll them one by one to a *sugeta* (wooden frame), taking care to make sure there aren't any air bubbles – this prevents the paper from peeling. I leave them overnight to allow the water to drain naturally.'

Then he slowly presses the sheets with a jack or stone weight, which squeezes out any remaining water and further binds the paper. Once this is done, they're peeled off the *sugeta* and stuck to a drying board and dried in direct sunlight. If the weather is bad, he uses a *dasuiki* (drying machine).

'Once all the materials are ready, making one or two sheets of paper isn't too difficult,' says Fuchikami-san.

'But to make 10 sheets of paper, or 100 sheets of paper? My concentration starts to go down. Fast-paced music, like heavy metal, really keeps me going!'

"

... to make 10 sheets of paper, or 100 sheets of paper? My concentration starts to go down. Fast-paced music, like heavy metal, really keeps me going!

CARING FOR YOUR WASHI PAPER

Keep washi paper in a cool dry place, away from bugs – especially the paper-eating kind like silverfish, cockroaches, termites and book lice. Framing your paper is the best way to do this, but you can also store it in a flat file or drawer – just be sure to keep small bug traps in the area.

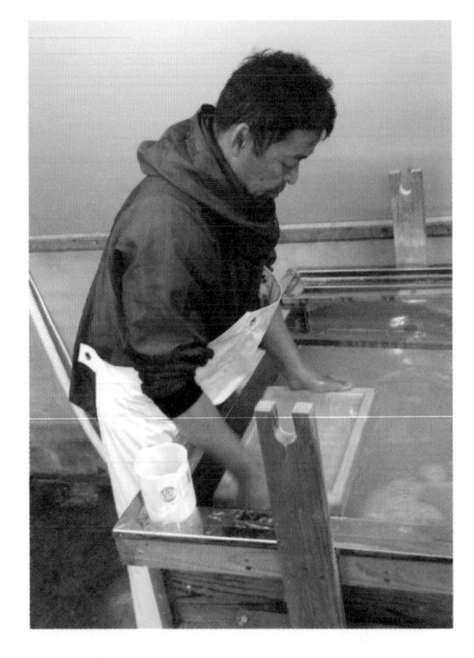

HOW TO

MAKE AN ORIGAMI MASU BOX

Traditionally, a masu box was a wooden box used to measure rice. A similar square, wooden box was also used to drink sake. In the origami world, a masu box is a traditional, square-shaped box made of paper.

MATERIALS

- 2 square pieces of washi paper (or other sturdy craft paper), one slightly larger than the other

Step 1

Step 2

Step 12

METHOD

1. Start with one square piece of paper. Fold the paper in half horizontally, then open it up again and fold it vertically.
2. Open the piece of paper up again and, with the patterned side of the paper face down, fold the four corners of the paper towards the centre point, to create a diamond shape.
3. Open up the triangular folds on the left- and right-hand sides.
4. Fold the top and bottom edges of the square into the horizontal midline, then open them out again.
5. Now rotate the model 90 degrees and open up the triangular folds on the left- and right-hand sides so that you have a square piece of paper again.
6. Fold the triangles at the top and bottom back into the centre line, then fold the top and bottom edges of the square into the horizontal midline. This time, leave them there.
7. Rotate the model another 90 degrees and then open out the first (square) folds; you should still have a triangle folded in on the left and another on the right.
8. Now lift the left- and right-hand sides up along the first fold on each side to form two of the sides of the box.
9. To form the third side of the box, pull the thumb and forefinger of each hand back towards your body until they are resting on the first folds out from the centre of the paper.
10. Now, with your forefingers on the inside of the folds' corners and your thumbs on the outside, raise the point of the paper closest to you and use your thumbs to gently push the paper inwards and forwards at the folds they're resting against, and the triangle upwards and over into the centre of the box.
11. To make the fourth side of the box, flip the model around and repeat steps 9 and 10 ... your box is finished.
12. To make a lid, make another box, but at steps 4–6, bring the folds 5 mm ($\frac{2}{10}$ in) out from the centre when you are making the folds that will form the sides of the lid.

UNIVE

Otto & Spike
AUSTRALIA

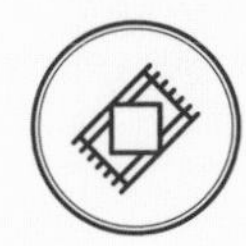

MCFADYEN PICNIC RUG

Machine knitting

Picnics have been around since at least the 1600s, when the French 'pique-nique' jumped the English Channel and became all the rage with the leisured classes. By the Victorian era, the common folk were starting to enjoy casual outdoor dining too. A top-quality picnic rug will last for years, and become a wonderful symbol of family togetherness. This one was made by the dedicated team at Otto & Spike in East Brunswick in Melbourne, who use surplus lambswool for all of their products. It's made using a robust jersey stitch, lined with a moisture-proof backing and comes with a carry handle. The rug was knitted on a Universal MC 728 Sinker machine that is more than 50 years old; its carry handle made on a 60-year-old strapping machine, using surplus cottons. The family patriarch, Les Mananov, doubles as the machines' mechanic.

MEASUREMENTS	1.6 × 1.6 m (5 × 5 ft)
MATERIALS	Australian lambswool, nylon, surplus cottons, yarn
KEY TOOLS	Dressmaker's scissors, latch hook, tailor's chalk
KEY MACHINES	Blanket stitch machine, overlocker, sinker machine, strapping machine
TIME TO MAKE	2 hours
LIFESPAN	Heirloom quality

LES, JIAN LING LIU, ANTHONY AND KIRSTIE MANANOV — Textile makers [Melbourne, Australia]

The Mananovs have been knitting under the moniker LMB Knitwear since 1969. Back when Les first started the business, East Brunswick was the centre of Melbourne's textile and garment industry. Today, very few companies knit in Australia. But Otto & Spike, a textiles label that the family created in 2006, is committed to designing and producing fine knitted goods locally.

'Everything is done in-house – everything, from the fixtures down to the string on our tags,' says Kirstie, who designs the textiles, marking things up stitch for stitch and pixel for pixel in Microsoft Paint.

'I know Paint is old school, but it works as it can be drilled down to that level of pixelation. The McFadyen is a thousand stitches wide, and a thousand stitches long. It's saved as a bitmap file then put into the Sinker's software, which translates it into stitches.'

Then Les and Anthony come on board, machine knitting each and every picnic rug. They have an extensive collection of mechanical and digital knitting machines, some of which are genuine antiques. Les purchased the one that the McFadyen is knitted on, a Universal MC 728 Sinker, when a neighbouring factory was closing down. It used to run on cassette tapes, but those are now obsolete, so it's connected to a laptop ... the size of a suitcase.

'The MC 728 was created in the early 1980s. It was one of the first coarse gauge (four gauge) knitting machines that could create a knit that replicated handknitting. It's capable of knitting up to eight different colours at any one time. It has a bed of 313 latch needles to knit with, and a series of weights and levers which mechanically manage the knitted fabric as it drops down from the machine.

'On top of the machine, there is space for 24 cones of yarn to be fed into the machine. One picnic blanket piece takes 36 minutes to knit. The machine itself is 4 m (13 ft) wide, 1.8 m (6 ft) tall and 2 m (6½ ft) deep. I hate to think how much she weighs!' says Les.

When the knitting is completed the rug is passed onto a machinist, who carefully sews the waterproof nylon backing onto the rug; first with an overlocker, to hold it in place, and then it's finished on a decorative blanket stitch machine. Ends are snipped with dressmaker's scissors; tailor's chalk and a latch hook, for picking up wayward threads, are also essential tools.

Les' career in knitting started when he got a job as a knitting machine mechanic at a knitting mill when he was 15 years old. He met his first wife (and Anthony's mother) there – she was working as a machinist. Seeing a future in knitting, Les saved his wages until he had enough to buy a single machine. He rented a shopfront and put the machine in the window with a hand-written sign offering commission knitting for fashion houses. Business grew and he soon had 20 machines and his own factory space.

Anthony joined the family business in 1990 at the age of 17, and is firmly at the helm today. Les' wife Jian Ling Liu joined the business as a steam presser in 1993, and Kirstie Mananov, Anthony's wife, joined as Otto & Spike's brand manager and designer in 1999. They have a small but hardworking team of 16 staff, and their ethically sourced and made beanies, scarves, blankets and picnic rugs are the toast of Melbourne.

'Over the years we had collectively managed to hoard an awful lot of surplus wool, and as knitting mills have closed their doors around us, we've bought their yarn,' explains Anthony.

Otto & Spike are also endorsed by Ethical Clothing Australia, which means their production processes are stringently audited.

'Sadly, being Australian made doesn't necessarily mean products are produced without exploitation. It's important to us to have this partnership, we want our customers to be conscious of where what they are wearing came from.'

CARING FOR YOUR MCFADYEN PICNIC RUG

After use, always give your woollen picnic rug a good shake and fold it up with the woollen side facing in. If it gets damp during your picnic, be sure to open it up and let it dry out when you get home. To clean the McFadyen, give it a lukewarm hand wash with an environmentally friendly laundry detergent and dry it flat in the shade.

"

One picnic blanket piece takes 36 minutes to knit. The machine itself is 4 m wide, 1.8 m tall and 2 m deep. I hate to think how much she weighs!

HOW TO

DARN A WOOLLEN RUG OR BLANKET

MATERIALS

- Rug or blanket to be darned
- Colour-matching yarn

TOOLS

- Scissors
- Darning needle
- Latch hook or crochet hook

METHOD

1. Locate the hole on your rug.
2. Tidy up the hole by cutting away any loose threads with a sharp pair of scissors.
3. Now carefully unravel the threads a little by gently worrying them with a darning needle to reveal a row of stitch loops at the top of the hole, and a row at the bottom.
4. Thread the darning needle with colour-matching yarn.
5. Beginning at either side of the uppermost edge, sew the stitches together in a criss-cross pattern, alternating from the top to bottom rows.
6. Once all the stitches are sewn, pull the yarn tight and tie it off with a knot at either end.
7. Using a latch or crochet hook, draw the loose thread ends back into the blanket fabric.
8. Your hole is repaired.

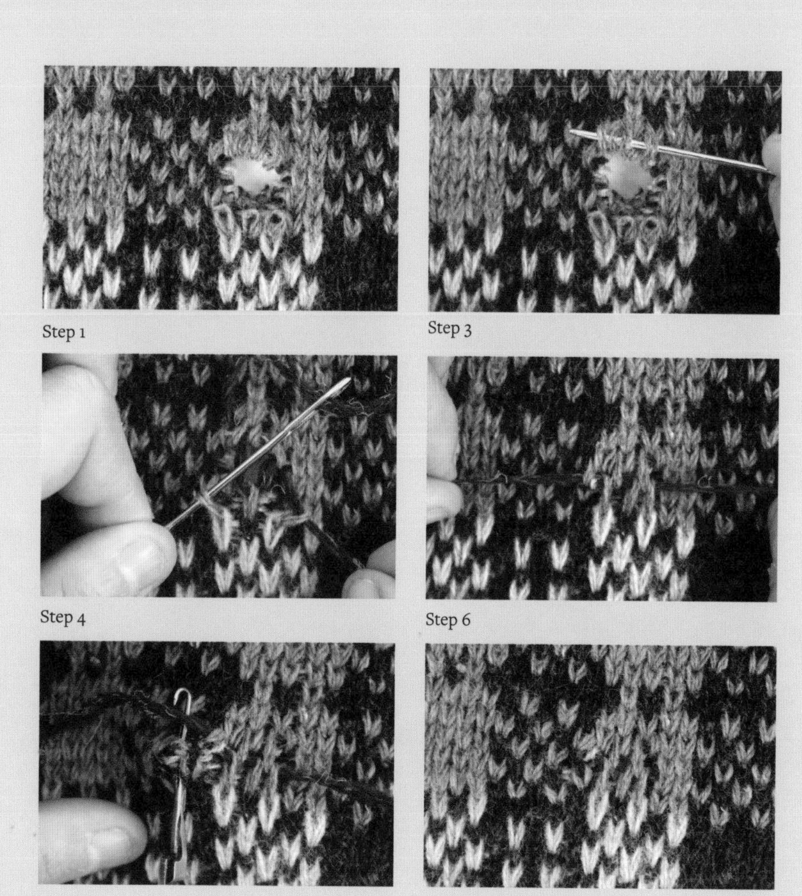

Step 1

Step 3

Step 4

Step 6

Step 7

Step 8

HANDMADE CONCRETE PLANTER

Concrete manufacturing

These planters bring a whole new meaning to the notion of a concrete jungle. Made entirely by hand, they make pretty and practical use of one of our era's most pervasive and durable materials: concrete. A composite of coarse aggregate (granular material such as sand, gravel, rocks or pebbles) bonded with fluid cement that hardens over time, concrete's history dates back to the Roman Empire, when it was used in structures like the Roman Pantheon, which was completed in 128 CE and still stands strong today. It's easy to feel lacklustre about concrete, but its versatility and texture make it perfect for creating long-lasting planters and homewares, as one clever Melbourne designer and maker has discovered. Kristy Tull's one of a kind planters come in two sizes and a range of colours that will likely change over time; all part of concrete's charm.

MEASUREMENTS	Small: height 8 cm × depth 9 cm (3 × 3½ in); medium: height 12 cm × depth 12 cm (4½ × 4½ in)
MATERIALS	Concrete, oxide, sealant, water
KEY TOOLS	Bowls, mixing tubs, sandpaper, spoons
MACHINES	Nil
TIME TO MAKE	7 days (including setting time)
LIFESPAN	20 years +

KRISTY TULL — Concrete homewares maker [Melbourne, Australia]

Kristy Tull began making jewellery from resin and polymer under the name Fox & Ramona in 2011. A couple of years later she discovered a faux concrete vessel in a junk store and loved its aesthetic and sturdiness. In true maker fashion, Kristy thought, 'I could make that!' – and that's where her love affair with concrete began.

The industrious maker had never made anything from concrete, but she didn't let that deter her. Kristy bought a bag of concrete mix and got to work in her back shed, experimenting with the medium.

'I've always loved craft and experimenting with different mediums but ultimately, it was experimenting with resin that led me to concrete. I found that there are many similarities in the making process, like the hands-on element and its unpredictable nature, so moving between the mediums was an easy process,' says Kristy.

That was back in 2013. Today, she works out of a rustic 150 sq m (1615 sq ft) shed in an industrial pocket of southwest Melbourne. She has a dedicated area for each phase – making, painting, sealing, packing and administration, along with two display areas where customers can check out her range of concrete planters, homewares and jewellery.

Each new project begins with design. Kristy decides on shapes, colours and patterns, taking inspiration from artwork and magazines. She sketches designs on paper, then adds colour.

'I don't play around with drawings for too long. I like to get in the workshop and start mixing colours to see if I can achieve the desired shade and colour combination while it's still fresh in my mind,' says Kristy.

Next, she creates reusable moulds out of silicone using measuring jugs and mixing bowls. 'I make about 20 of each size planter mould at a time, so I can pour multiple planters.'

Now it's time to make the planters themselves. Kristy uses a cup to measure the concrete powder and a spoon to measure the oxide, pouring them carefully into a large plastic bowl. The ratio of concrete to oxide differs depending on the colour Kristy is trying to achieve; each of these planters takes approximately one cup of concrete, and one tablespoon of oxide.

She puts on latex gloves, then mixes the concrete with water using a stainless steel spoon. She spoons the concrete into the moulds, then leaves it to set.

Between 24 and 48 hours later, it's time to de-mould the planters. Kristy does this by peeling off the silicone mould. Each mould is opaque, so it's only now that Kristy can see how close her concrete mixing and pouring is to her concept design. Each planter turns out differently.

'The finish of each planter is unpredictable –no two pieces ever look the same. It's a long 24 to 48 hours setting time before I can de-mould, especially when working on new designs. I love discovering the unique colour swirls and small air bubbles that give each piece its own identity.'

Next, Kristy smooths the surface of each planter by sanding it with 1200 grit sandpaper, then washes it under cold running water to remove any dust. She puts the planters on a drying rack, where they dry for up to five days, depending on the weather.

When the planters are completely dry, Kristy applies a concrete sealant to the interior and exterior using a paintbrush. This makes the planter waterproof and stain resistant, and gives it a light sheen finish.

'I feel extremely lucky to be designing, making and creating as my job. I get a great sense of achievement when making something from scratch using only my hands, a bowl and spoon.'

CARING FOR YOUR HANDMADE CONCRETE PLANTER

These sturdy planters can be used both in and out of doors, but they'll stay looking their best if kept inside, away from the elements. If they get dirty, simply wipe them with a damp cloth.

"

I get a great sense of achievement when making something from scratch using only my hands, a bowl and spoon.

HOW TO

MAKE A CONCRETE PLANTER

MATERIALS

- Cement and aggregate mix
- Water

TOOLS

- Metal spoon
- Measuring cup
- Bucket
- Two nesting vessels to form your mould. Old yoghurt containers or plastic planters work well – just make sure one fits inside the other, with a bit of a gap.
- Cooking oil, preferably in spray form
- Rocks
- Pliers (optional)
- Sandpaper

PRO TIPS

- *If you want aesthetic air pockets in your planter, avoid vibrating (patting, tapping, shaking or banging) your cement mixture down too much after placing it in the mould. If you intend to use your planter to house plants that require drainage, use a masonry bit to drill a hole or holes in the base.*
- *Concrete is porous. Any moisture it comes into contact with will seep through and dampen the base of your planter. Protect any underlying surfaces by attaching small felt or rubber pads to the base. You may even want to coat the interior with a waterproof concrete sealer, available from hardware stores.*
- *In addition to coating the interior of your planter, you can also paint the exterior with a clear sealer to give it a slightly smoother feel and a hint of sheen.*

METHOD

1. Combine your cement and aggregate mix with water according to the instructions for the type of cement you're using. Place a cup or two (or three or four) of cement in your bucket, then gradually add water, stirring as you go. You're after a workable, toothpaste-like consistency, so don't be afraid to add more water if it seems too dry, or more cement if it's too wet.
2. The amount of cement you'll need will depend on the size of your mould – one of Kristy's small planters uses around one cup of cement.
3. Lubricate your moulds by generously coating the two surfaces that will be in contact with the cement mixture with the oil spray.
4. Check the depth you'll need your inner mould to sit at by holding it at your desired height and taking note of where it sits in relation to the rim of your outer mould. This will help to ensure your base isn't too thin or too thick.
5. Now use the spoon to fill your outer mould with cement mixture. Ensure you leave enough space for your inner mould to take up some of the volume.
6. Press your inner mould down into the cement mixture. Take care to keep it as level and central as possible, and don't push it down too far – you need the base of your planter to be around 1.5 cm (½ in) thick. If there is too much or too little concrete in your mould, simply add or remove some.
7. The inner mould will want to float up, so weigh it down by filling it with rocks (spare aggregate works fine) or anchoring it with some masking tape.
8. Leave it to cure in a protected area for around 24 hours.
9. After 24 hours, de-mould by taking out the inner mould and removing the planter from the outer mould. Pliers may come in handy here.
10. Leave the planter somewhere out of direct sunlight for a week. This will help it to cure well, and avoid cracks and breakage.
11. Sand or file any rough areas, if desired.

EBB TIDE PEBBLE PLATES

Ceramics

Ceramics are some of the oldest human artefacts ever found. Kiln-fired animal and human figurines made from clay and water dating back to 24,000 BCE have been discovered, while pots for water and food storage date back to 10,000 BCE. The Egyptians invented glazes for platters and the like not long after. This pebble-shaped range of plates, Ebb Tide, has been handmade by ceramicist Kim Wallace, who was born and raised in the Netherlands, but has called Australia home since 2000. Raw and organic, Ebb Tide draws on colours and patterns from nature and treasures left on the beach by the receding tide. Each one-of-a-kind piece is made by hand from porcelain clay, handpainted with a matte glaze, and has an unglazed base and edges with a vitrified stone-like surface. Like the pots of the ancients, they're fired at extreme heat, which makes them durable and incredibly long lasting.

MEASUREMENTS	Dinner plate: 26 × 24 cm (10 × 9½ in); bread plate: 20 × 19 cm (8 × 7½ in); dessert/side plate: 18 × 16 cm (7 × 6 ½ in); small plate: 14 × 12 cm (5½ × 4½ in); tiny dish: 7 × 5 cm (3 × 2 in)
MATERIALS	Australian porcelain clay, glaze
KEY TOOLS	Hake brushes, knife, polymer ribs, sponges, wooden board
KEY MACHINES	Kiln, slab roller
TIME TO MAKE	1 week (including drying and firing time)
LIFESPAN	Heirloom quality

KIM WALLACE — Ceramicist [Sunshine Coast, Australia]

In 2008, Kim Wallace was taking a break from her job as a graphic designer when she started playing around with clay. She found a box full of gorgeous, hand crocheted doilies in a second-hand shop, and experimented with pressing the patterns into the clay. Then she drew on her graphic design skills and began creating interesting compositions and colourways.

'I started putting my creations out there at markets and in my online shop, and to my big surprise they sold quite well. I've never looked back,' says Kim, who now runs her own home-based ceramics studio in the Noosa Hinterland on Australia's Sunshine Coast, and supplies some of Australia's leading restaurants with tableware.

Kim's studio is a purpose-built artist's studio surrounded by rainforest. It's located above her children's bedrooms in the midst of an 8-acre (3 ha) property, and it's the reason she and her husband, Greg, moved into the place.

'It was previously used by a painter and there are paint splatters all over the gorgeous timber floors, high ceilings and wooden beams. I love being surrounded by the native wildlife, cockatoos and kookaburras making noise around us and the beautiful breezes flowing through.'

When Kim is ready to start working on a new batch of plates, she begins by cutting clay into slabs and rolling it in a slab roller. She smooths the top with a polymer rib (a flat, kidney shaped tool with rounded edges that fits in the palm of the hand), one piece of clay at a time. Then she uses a knife to cut out the shape of the plate, and transfers it to a wooden board to dry a little. She smooths and rounds the edges with her hands, then gently turns it upside down and imprints her KW Ceramics logo on the base.

'Next I place a plaster mould on the base, and shape the edges of the plate by pressing the clay into the mould and smoothing it,

again with my hands. My hands are my most important tools.

'I transfer the plate to a board again, and leave it to dry for a few days, depending on the weather. I check the plates several times to ensure even drying and no warping. Once it's dry, I sponge it with room temperature water; this makes it nice and smooth all over.'

After sponging, Kim leaves the plates to dry for around 30 minutes, then glazes each plate by hand, applying several coats of glaze with a hake brush, an oriental-style wash brush with a long flat handle.

'I love a good brush, as a painter would. It's a vital tool to getting the look of the ceramic piece right. My ceramic finishes are simple in nature so the application is most important. Even something like the sponges we use for smoothing the piece can make a huge difference in how the process feels.'

When Kim has a kiln's worth of plates ready to fire (around twelve plates with smaller pieces fitted alongside them in the kiln), she places them in her electric kiln, and waits. The kiln fires at 1280°C (2336°F) for 13 hours, and takes 24 hours to cool slowly.

'Unloading the kiln is always exciting; hoping all pieces have fired nicely and there are no major disasters. But it can be nerve-racking too, especially when I'm trialling new work.'

When the plates have cooled, she removes them from the kiln and sands their unglazed surfaces to give them a smooth feel; a subtle touch that makes them lovely to handle.

'When making ceramics you have to be in the moment. If your mind is elsewhere then it most often reflects in breakages or cracking. I love working with my hands, it's very grounding and calming to be in my own space for a while. It's also incredibly satisfying making functional pieces that people love from a humble block of mud.'

CARING FOR YOUR CERAMICS

Gently handwashing your ceramics will increase their longevity. Take care to avoid sudden temperature changes, such as taking a plate from a hot oven to a cold benchtop, as this can cause thermal shock, which can crack the piece. To prevent scratches to the glaze, store your ceramics in plate racks or with a soft cloth between each piece. To keep your pieces beautiful, avoid unglazed surfaces coming into contact with staining foods or liquids.

"

I love working with my hands, it's very grounding and calming to be in my own space for a while. It's also incredibly satisfying making functional pieces that people love from a humble block of mud.

HOW TO MAKE

A CLAY SPOON

MATERIALS

- Porcelain clay
- Jar of water
- Underglaze (optional)
- Glaze

TOOLS

- 2 sheets of heavy duty canvas
- Slab roller or rolling pin
- Polymer rib
- Pin tool
- Spoon template (optional)
- Stamps (optional)
- Wet sponge
- Kiln

PRO TIP

All white or light-coloured tableware, whether store-bought or handmade, can develop grey scuff marks over time from cutlery use. These can be easily removed by rubbing a paste of bicarbonate of soda (baking soda) and a little water over the marks with a cloth.

METHOD

1. Place your clay between two layers of canvas.
2. Roll your clay to around 5 mm (⅕ in) thick using a slab roller or rolling pin. Remove the top layer of canvas.
3. Wet the polymer rib, then use it to smooth the surface of the clay. This compresses the clay and gets rid of any texture left by the canvas during rolling.
4. Cut your spoon shape using a pin tool. Kim likes using a template (cut out of card or plastic) to ensure her spoons are the same size.
5. Gently remove your spoon shapes from the clay and put them aside on the canvas to dry for 5–15 minutes (how long depends on the wetness of the clay and the weather).
6. When the spoons seem ready to be handled without losing their shape, use your finger to smooth and round off all edges of the spoon.
7. Stamp your design onto the spoon and your maker's mark on the reverse, if desired.
8. Now it's time to shape the spoon. Place the bowl of the spoon in the palm of your hand and press into the bowl with your thumb to create a nice shape.
9. Rotate your thumb and press again on the other side of the bowl until you are happy. Kim likes to pinch the base of the bowl of the spoon to make the edges a little thinner than the rest of the spoon.
10. Use your thumb and forefinger to curve the handle into your desired shape.
11. Let your spoon dry on a timber or other slightly absorbent surface for a day.
12. Once dry, sponge the spoon with a wet sponge to smooth off all the edges.
13. The spoon can now be coloured, glazed and fired according to the instructions for your glaze and kiln.

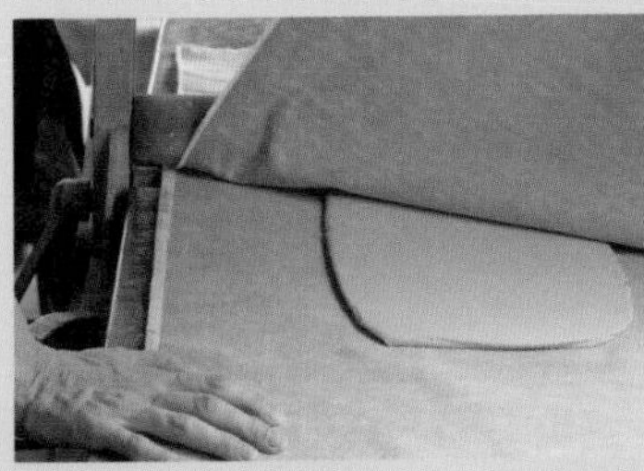
Step 2

Step 5

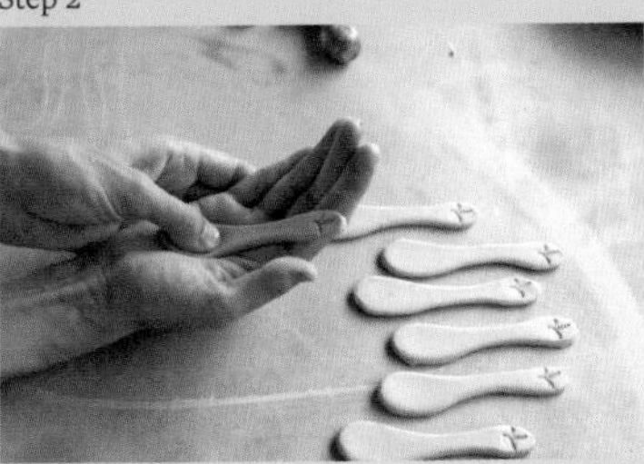
Step 8

Step 12

Connor
Daniel
Zenith Records

REGURGITATOR
UNIT
YOU AM I's #4 RECORD
urthboy
WHITE CAVES
BUTTERCUP RECORDS

7" AND 12" VINYL RECORDS

Vinyl pressing

Vinyl is arguably the most persistent musical medium of our age. The first records were printed onto shellac in 1926, used for playing motion picture soundtracks. By the mid-1930s, all records were being pressed onto a new type of plastic – polyvinyl chloride (PVC) resin, or vinyl. When Columbia Records introduced the 10- and 12-inch LP – long play – records in 1948, they were swiftly adopted as standard by the entire record industry. Today, the technology for vinyl pressing and playing remains remarkably similar to the way it was in 1948. These records are proudly made at Australia's only remaining vinyl pressing plant, in a space that was once a bona fide sweatshop in the backblocks of Brunswick, in Melbourne. Zenith Records press vinyl in 7- and 12-inch formats on equipment that dates back to vinyl's heyday in the 1960s and 1970s. Oh, and they do colour. Lots of colour.

MEASUREMENTS	Diameter: 17.5 cm (7 in) or 30 cm (12 in)
MATERIALS	PVC, printed paper labels
KEY TOOLS	Allen keys, copper head hammer hex, jaw shifting spanner, multimeter, screwdrivers, soldering iron, spanners (wrenches)
KEY MACHINES	Automatic presses, boiler, cooling tower, edge trimmers, extruders (electric and steam), finishing-off baths, manual press machines, Neumann lathe, pre-plating baths
TIME TO MAKE	8 hours
LIFESPAN	1 year +

CONNOR DALTON AND DANIEL HALLPIKE — Cutting engineer and phonographic engineer [Melbourne, Australia]

Connor Dalton and Daniel Hallpike are two crucial members in Zenith's eight-strong team. As cutting engineer, Connor transfers digital music files onto lacquers via a 1974 Neumann lathe. As phonographic engineer, Daniel's job is to ensure the smooth running and maintenance of the company's three manual press machines.

The first step in making records is transferring the audio into physical grooves on an aluminium disc coated in nitrocellulose known as a 'lacquer'. It's vital that this is done properly, as it determines the sound.

The two lacquers (one for each side) are carefully cut using a 1974 Neumann cutting lathe by Zenith's cutting engineer, and eventually form the mould for the pressing stage.

Next, Zenith's plating team make nickel stampers from the cut lacquers. The soft lacquers are electroplated in nickel plating baths, causing a negative (a 'stamper') to develop from the nickel solution to an exact weight and thickness. It's this, the nickel stamper, that the records are pressed from.

Plating involves silvering the lacquer (creating a conductive face so that the nickel can be electroplated); plating the silvered lacquer (electroplating and building up sufficient nickel to make a stamper); and passivating the stamper and replating it all over again so that a metal 'mother' – a playable, positive metal plate – can be created. The stampers then need to be sanded, polished, centre punched and 'coined'.

'We make the mother so that further stampers can be made for larger runs, or in case the plate is damaged on the press. Coining the stamper presses the centres and outer edges to the profile of the moulds. This is what they're mounted against on the press,' says Connor. Next, it's time to prepare the paper labels for records.

'We print sets of labels on large sheets, which are die cut then centre drilled. We bake them for at least 24 hours so that the moisture in the paper is removed, as labels that have not been sufficiently dried will cause issues on the press.'

Then it's time to press vinyl. The press operator carefully mounts the plates to the moulds on the press. The moulds are a highly polished fixture on the press with cavities that allow steam and water to flow through, heating then cooling the pressed vinyl. Each press has an accompanying PCV extruder, which heats the PVC pellets into a malleable, plastic form.

'The extruder delivers an exact amount of heated material into what we call a pattie. Once that's happened, we manually place the labels on centre pins in the press, then place the pattie at the centre of the bottom mould. The operator closes the safety gate, which initiates the press. A pre-heating cycle heats the moulds and then a second heat cycle throttles the steam to keep it constant, as the press exerts 120 tonnes of pressure, which allows the stampers to fully penetrate the PVC and press the grooves.'

After that, a cooling cycle sets the PVC, and the record can be removed as a warm but fully formed rigid disc. The crew then place the pressed record on an Alpha Toolex cutting machine and trim the excess vinyl to create a clean, round edge. It's then placed in an inner sleeve and stacked on a spindle; as more records are pressed, sleeved and stacked, aluminium spacers are placed in-between to ensure they cool adequately.

Finally, the records are packed into printed sleeves, shrink-wrapped and boxed for dispatch.

This sounds straightforward, but the pre-1980s vintage of the machines Zenith works with means Daniel needs to be on his game every day of the week. His toolkit, which contains Allen keys, a copper head hammer hex and jaw shifting spanner, multimeter, screwdrivers, soldering iron and spanners (wrenches), is always close to hand.

'When the plant is running like clockwork it's very satisfying, but the age and nature of the machinery means that if one thing goes down, nine times out of ten we'll get a steady stream of breakdowns, which can quickly make us feel in over our heads. I spend my time problem solving and sourcing or making hard-to-replace parts. Many of the faults that arise aren't easily identifiable; we have to stop, assess and think. When we're pressing records, the temperature in the workshop automatically goes up!'

CARING FOR YOUR VINYL RECORDS

A record's lifespan is potentially infinite; how long it lasts really depends on the care it's given. To make your vinyl last as long as possible, always put the records back in their inner sleeves and covers after use to avoid exposure to dust and dirt particles. Handle them with care. Only hold records by the edges, and the centre labels, and avoid leaving fingerprints on the grooves. To remove dust particles before and after playing the record, brush it with a carbon fibre brush. Keep records stored in a vertical position, never flat, or they will warp. Another warp alert – take care not to leave them in the sun or expose them to extreme temperatures.

“

I spend my time problem solving and sourcing or making hard to replace parts ... When we're pressing records, the temperature in the workshop automatically goes up!

HOW TO

MAKE A SPLATTER RECORD

Splatter records are records with colourful splatters on the vinyl. You won't be able to try this at home unless you have a vinyl pressing plant in your basement; here's how Zenith do it.

MATERIALS

- PVC vinyl – base colour
- PVC vinyl – splatter colour
- Paper labels
- Inner sleeves

TOOLS

- Spindle
- Aluminium spacers

MACHINES

- Steam extruder
- Electric frying pan
- Alpha Toolex manual press
- Alpha Toolex cutting machine

SAFETY GEAR

- Gloves

METHOD

1. To create a splatter record, the first thing Zenith's press operators need to do is heat up the PVC of both the base colour and the splatter colour vinyl.
2. They pour the base colour PVC into the extruder in pellet form, where it is heated and extruded into what's known as a 'pattie'.
3. They heat the splatter colour, in this case white, in an electric frying pan.
4. When the base colour vinyl pattie is extruded, the press operators place it in the frying pan of heated white vinyl, where the pellets are sprinkled and pressed firmly, so that they stick to the pattie. Then they flip it over and repeat the process on the other side.
5. They manually place the labels on centre pins in the press.
6. They place the pattie at the centre of the bottom mould, and the operator closes the safety gate to initiate the press. A pre-heating cycle heats the moulds and then a second heat cycle throttles the steam to keep it constant as the press exerts 120 tonnes of pressure, which allows the stampers to fully penetrate the PVC and press the grooves. A cooling cycle then takes place, which 'sets' the PVC and allows the record to be removed as a warm but fully formed rigid disc.
7. They put the pressed records onto an Alpha Toolex cutting machine and trim the excess to create a clean, round edge.
8. They place the records in inner sleeves and stack them on spindles with aluminium spacers between them so they can adequately cool.
9. Now the splatter records are packaged up, ready for dispatch.

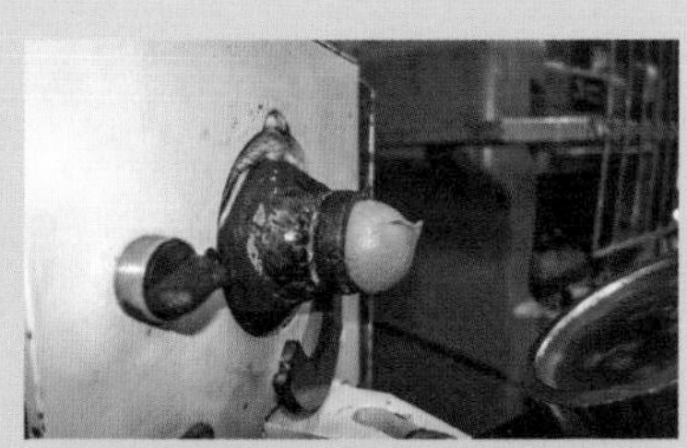

Step 2

Step 3

Step 6

WE PRESS
VINYL
WWW.ZENITHRECORDS.ORG
FOR QUOTES CALL:
(03) 9311 0075
OR EMAIL:
INFO@ZENITHRECORDS.ORG

INTERVAL
HEATING 1
HEATING 2
COOLING
EJECTION
SECONDS
SECONDS H1
SECONDS H2
SECONDS C
SECONDS E
GATE
TEMP CYCLE
OPEN
CLOSED
PRESS CYCLE
OUTLET
MIN.
MAX.
FUSE1A
FUSE1A
COOL
AUTO
A
B

DAPPLE-GREY STALLION ROCKING HORSE

Toy making

Ah, the rocking horse, apple of any child's eye. Chances are you've ridden one. If you were lucky, maybe you had one growing up. It might even have been your mother's or father's before you – these beauties are saddled with the 'family heirloom' stamp. This particular steed is a dapple-grey stallion, made by rocking-horse maker and restoration expert Olivia O'Connor. Olivia made it in a workshop on a farm with real horses, nestled in the rolling hills of Berry's Creek in South Gippsland, Australia. No two of Olivia's rocking horses are alike, but they're all hand carved from sustainably sourced timber and finished with real, ethically sourced horsehair and tough, hardwearing lacquers. Olivia makes the saddles, too, using the finest Australian leathers. They come complete with stirrups, girth, bridle and bit. Giddy up!

MEASUREMENTS	Height: 1.1 m (3 ½ ft)/made to measure
MATERIALS	Horsehair, lacquer, leather, timber
KEY TOOLS	Chisel, flexible drawknife, standard drawknife, gouge, rasps
KEY MACHINES	Band saw
TIME TO MAKE	30 days
LIFESPAN	Heirloom quality

OLIVIA O'CONNOR — Rocking horse maker [South Gippsland, Australia]

The first time Olivia O'Connor made a rocking horse, she had to carve the head twice. The horse was a bay stallion, and it was her final project at the National Institute of Dramatic Art in Sydney, where she had spent three years studying prop making.

Students were given the time and the budget to make anything they wanted. For Olivia, it was a close call between an animatronic pig's head and a rocking horse. 'I grew up on a farm, and my father was a horse trainer. I loved riding when I was younger, but I never had a rocking horse as a child ... I'd secretly always wanted one, so the rocking horse won out.'

'Of all the horses I've made, the first was the most difficult. Apart from liking them, I didn't actually know much about rocking horses. I had to research their design history and decide what style of horse to make. Then I drafted all the patterns and made the horse from scratch. The first head I made was horrendous! I had to carve it again.'

That first rocking horse was in the style of British rocking horses typical of the 1860s. 'They're quite realistic. They were making rocking horses around that time in Germany, too, but theirs were quite severe and angular. Then in the 1940s you get quite unreal-looking horses; ones with crazy nostrils and skinny necks.'

This clever craftsperson, whose experience as a prop maker and scenic artist includes a stint as a saddle maker for the New York stage production of 'War Horse', especially liked that the heads of the older, British horses turned slightly to the right. This subtle asymmetry shows that the horses haven't been carved by a machine, and is a feature Olivia retains in her own rocking horses.

Since making that first bay stallion, Olivia has made more than 40 rocking horses and restored almost as many antique steeds and vintage nags. She can pick the age and make of a rocking horse just by looking at it. All of them, whether they're hand or factory made, have systematic weak spots. Olivia has learned them all and takes care to avoid them in her own creations.

These days, Olivia's horses come in three sizes: stallion, mare and pony. When a new commission comes in, she gets started by carefully selecting the timber, checking for cleanness, straightness of grain and no knots or weaknesses. Next, she cuts the timber and laminates a hollow, internally-braced box together; this forms the body of the horse. It's carved using a flexible drawknife – that's a curved and flexible blade with a handle at either end – and rasps.

Then Olivia laminates the timber for the head and neck together and cuts out the shape of the horse's neck on a band saw. The neck is attached to the body using hidden joinery and is shaped using a gouge, chisels and rasps. Next come the legs, then the head. Olivia allows plenty of time for carving the head, as it contains the most detail and has to be 'just right'. It's attached to the neck with more finely hidden joinery, then the entire body is sanded and prepped for painting.

Next comes saddle making, including dying the leathers and carefully placing hundreds of stitches, making and attaching the tail, mane and eyes, making and oiling the stand and, finally, assembling the horse. From start to finish, the process takes an entire month.

'It's really sweet when I drop off a horse at someone's house, or they come and collect it and the children are there. They think it's the most magical thing in the world, and that's special, because I love knowing that something I've made is going to be looked after and cherished.'

"

I love knowing that something I've made is going to be looked after and cherished.

CARING FOR YOUR ROCKING HORSE

If treated well, a rocking horse will last for hundreds of years and make several generations of children very, very happy.

One of Olivia's sturdy horses can comfortably take the weight of an adult rider up to 80 kg (175 lb). An antique or vintage horse in good condition can also take a fair bit of weight.

Rocking horses need to be kept clean and dry. They should be kept out of direct sunlight and away from sources of heat and cooling such as heaters, open fires and air-conditioning, and wiped down occasionally with a clean, dry cloth.

If the mane becomes dirty or tangled, just treat it as you would a normal head of hair, and give it a gentle shampoo and comb. Just make sure you protect the rest of the horse by covering it with towels, so it doesn't get wet.

You might even like to encourage the budding equestrians in your household to cover the horse with a blanket in between rides.

If you need to store a rocking horse, don't put it out in the garden shed! It will become damp and rot. And don't be tempted to sit boxes or other items on its back; over time, these will weaken its structure. Instead, wrap it in a clean blanket and put it somewhere warm and dry, such as an attic or a storage facility.

HOW TO

MAKE A HOBBY HORSE

Okay, so this is no rocking horse, but a simple hobby horse will keep the kids happy for hours too.

MATERIALS

- 19 mm (¾ in) plywood
- 950 mm (37 in) length of 25 mm (1 in) thick hardwood dowel
- 250 mm (10 in) length of 25 mm (1 in) thick hardwood dowel
- 120 mm (5 in) length of 45 mm (1⅗ in) thick hardwood dowel
- 100 mm (4 in) of 9.5 mm (⅖ in) thick hardwood dowel
- PVA glue (ideally with nozzle)
- Leather or canvas
- Timber oil
- 6 × 15 mm (⅗ in) long Phillips head, 8 gauge screws
- 4 × upholstery tacks

TOOLS

- Horse head pattern – trace one from a picture. Here Olivia's head measures 34 cm (13 ⅖ in) from the top of the head to the base of the neck, and 40 cm (16 in) wide
- Pencil
- Jigsaw or band saw
- Drill
- 25 mm (1 in) spade drill bit
- Handsaw
- Coping saw
- Damp rag
- Sandpaper
- 2.5 mm (9/100 in) drill bit
- 9.5 mm (7/20 in) drill bit
- Driving bit for drill or screwdriver
- Sharp scissors or leather knife
- Hammer

METHOD

1. Trace the horse head pattern (or draw one freehand) onto the plywood.
2. Cut out the shape using a jigsaw or a band saw.
3. Drill a hole for the handle in the neck using the spade drill bit.
4. In one end of the 45 mm (1⅗ in) hardwood dowel, drill a 50 mm (2 in) deep hole using the spade drill bit.
5. In the other end, mark a centre line 50 mm (2 in) down the length of the shaft. Measure 9.5 mm (⅖ in) from this line on each side and mark new lines. This will be the slot that the horse's head will slide into.
6. Using your handsaw, carefully cut down the two outer lines.
7. Next, use your coping saw to horizontally cut out the base of the slot.
8. Coat the inside of the slot with PVA glue and slide the horse's head into position, taking care to wipe excess glue off with a damp rag.
9. Using the 2.5 mm (9/100 in) drill bit, drill through the dowel and into the plywood for the eyes and handle.
10. Change to the 9.5 mm (7/20 in) drill bit and countersink the top of your pilot holes. Do two holes on the first side and one on the second side.
11. Drive your screws into the holes. Hammer two upholstery tacks for the eyes.
12. Plug the tops of the screws with glue, then firmly insert the 9.5 mm (7/20 in) dowel.
13. Repeat steps 9–12. This time glue and drill through your larger hardwood dowel 'collar' and into the longer piece of dowel.
14. Sand and oil the hobby horse and handle.
15. Use your scissors or leather knife to cut out two ears from the leather or canvas.
16. Fold the base and nail one ear to each side of the head using the hammer and upholstery tacks.
17. Hammer the handle into the hole and get ready to gallop away.

PRO TIP

- *When oiling your hobby horse, allow the oil to sink into the timber, then give it a light sand with 320 grit sandpaper, then another light oil, and rub with a clean cloth. Using oil instead of varnish means that if your hobby horse gets knocked, you can sand the damaged area and re-oil it to have it looking new again. Organoil is food safe, and therefore perfect for children's toys.*

Step 1

Step 9

Step 10

Step 11

Step 13

Step 14

Step 16

SHELL CORDOVANS

Shoe making

There are shoes, and then there are shoes. And these shoes, my friend, have been handcrafted in a small workshop in Seville, Spain, by a leatherworker who only makes 60 pairs a year. These are Antonio Enrile's Shell Cordovans. They're made from 100 per cent lubricated, vegetable-tanned horsehide, sourced from Horween in the United States (one of the most reputable tanneries in the world) and used flesh-side out. Lubrication provides flexibility, while the vegetable tanning means the leather is breathable, and will improve with age. Unlike shoes that are machine or factory made, Antonio's shoes are hand-welted. A handcrafted welt is a major point of difference between traditional and modern, machine-based shoe making, and one that makes Antonio's shoes incredibly robust. It also means that the sole can be replaced time and time again, so his shoes can last decades.

MEASUREMENTS	Made to order
MATERIALS	Aniline dye, beeswax, carnauva wax, cork, leather, natural flax, vegetable-tanned Genuine Shell Cordovan leather, wood
KEY TOOLS	Awl pricker, glass, hammer, key (shoemaker's knife)
KEY MACHINES	Finishing machine (to shape the heel), sewing machine (for sewing the upper)
TIME TO MAKE	4 days to make, + 20 days drying
LIFESPAN	Heirloom quality

ANTONIO GARCIA ENRILE — Shoemaker [Seville, Spain]

Antonio Enrile first fell in love with handmade leather shoes when he was twelve. A couple of years later, he started making belts out of leather; soon, his heart was set on becoming a craftsman. Today, he hand makes shoes and other leather goods in his workshop, Enrile, in Seville, and sells them from his store, out front.

Antonio says there are no secrets to making shoes. His method echoes that of shoemakers from Elda, one of Spain's most traditional shoemaking cities, and the place where he studied his craft. Like the bespoke shoes of old, Enrile shoes are made almost entirely by hand.

After Antonio has measured his client's foot and they've agreed on a shoe design, he uses a key (shoemaker's knife) to cut the leather, wets it, and puts it on the last (wooden foot mould). He gives the leather a firm hammering; as it dries, it shrinks to fit the last.

Next comes construction of the internal supports. It's a fiddly business that involves working a few steps ahead to ensure the pieces will fit perfectly once they're pieced together. The insole (the stuff between the outsole and your feet) is filled with cork, and a piece of wood called the shank is placed inside, for stability.

Then the insole is prepared for the construction of the welt, which is a strip of leather that runs along the perimeter of the insole (the part that touches your feet) and is sewed onto the upper (the part that surrounds your feet) and to the outsole (the part that touches the ground).

The insole and outsole are glued together and adjusted to fit the welt, then they're wet and a channel is cut into the leather sole, where the seam of the welt will go. The leather is wet again, and then Antonio uses his favourite tool, a Blanchard round awl pricker, to make the holes for the thread.

'It's made from wood and bronze, and is more than 100 years old. It has all the charm of a well-made tool. It was probably used for saddle making; using it would have meant all the work that came out of that workshop would have had the same stitch, regardless of who carried out the work.'

Antonio takes care to make the holes exactly the same size as the thread, so that when the thread crosses over, the holes won't be noticeable. He twists several threads together then uses a cross-sewing technique to attach the sole, welt and upper together.

'Each stitch needs to be tightened up and independent, and is carefully hidden away in the channel. This means that even if the thread wears down, the sole will retain its appearance.'

The hand welting is the most physically demanding of all the stages, and one that Antonio takes great pride in doing thoroughly. He finishes by using a tool to perforate each stitch of the welt. It's not strictly necessary, but it looks good.

The final stages of shoe construction involve making the heel. Unless the client's needs demand otherwise, it's vital to the comfort of the shoes that the levels are perfectly matched. The edges of the sole and the sole itself are refined with glassing (scraping with a thin piece of glass to remove the top layer of the leather) and then light sanding, then they are painted. When the paint is dry, Antonio applies a layer of beeswax. The edges and soles are burnished with hot iron bars to seal the paint, then they're polished with a cloth. The uppers are nourished and polished with creams and waxes, and then the shoe is left to dry on the last for at least 20 days.

Antonio likes to get into his 50 sq m (538 sq ft) workshop early in the morning. This is when he feels most productive, and is able to lose himself in his craft, before the demands of the business day descend.

'In the beginning I was a craftsman; that led to me being a businessman. But my soul will always be a craftsman's soul.'

"

In the beginning I was a craftsman; that led to me being a businessman. But my soul will always be a craftsman's soul.

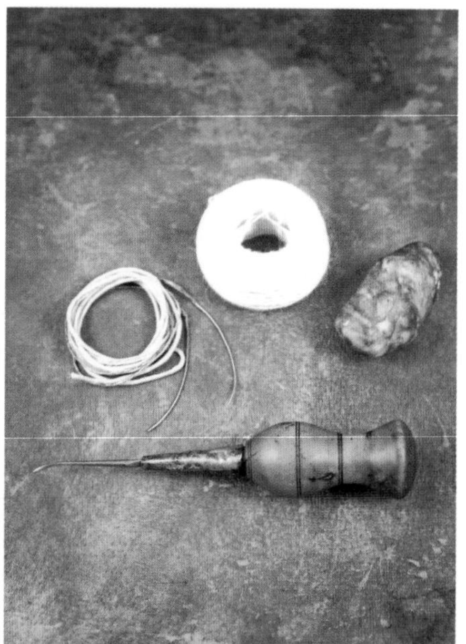

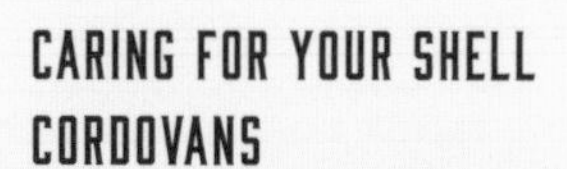

CARING FOR YOUR SHELL CORDOVANS

Clean your Shell Cordovans by wiping the uppers with a damp cloth and brushing them gently with a polishing brush. Because the leather is lubricated and tanned with vegetable oils, they'll never need (or tolerate) coloured waxes or creams. But do hydrate them with a high-quality, colour-neutral wax cream.

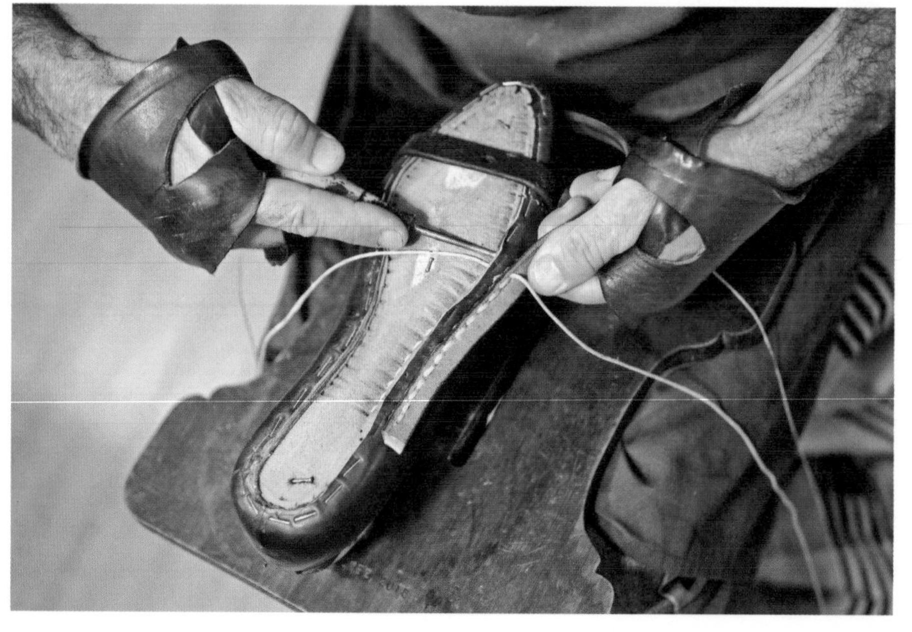

HOW TO

MAKE YOUR SHOES LAST AS LONG AS POSSIBLE

Rotation – feet are humid, sweaty little beasts. Give your shoes a break by never wearing the same pair for more than two days running.

Keeping the shape – after wear, place shoe trees inside the shoes to help them keep their shape.

Drying – if your shoes get wet, let them air dry on their sides (so the soles can also dry); never put them in or near a heat source, as this will dehydrate the leather.

Cleaning – always put shoe trees in your shoes before cleaning them.

Replacing parts – replace the toecaps and soles as soon as needed; letting this go for too long is not healthy for your shoe, or your foot.

Using the right brush – experts recommend using three different brushes, in this order, for tiptop results:

1. **Dirt removal** – horsehair brushes are soft and smooth, and will gently remove dirt and mud from leather uppers. If the shoes are very dirty, you can dip the brush in a little water.
2. **Dauber** – for applying shoe cream; sure, you could use a cloth, but a dauber brush, gently rotated over the uppers, will really get the cream in the there.
3. **Polishing** – brushes crafted from Cashmere goat hair can produce exceptional shine in leather.

Leave the shoes to rest for at least 20 minutes between each stage.

If you're wearing suede or nubuck, you'll need a crepe rubber or brass brush to clean your delicate footwear. Take care to only brush in the direction of the grain.

Never use a synthetic brush. They might be easy on your pocket, but they're hard on your shoes. This is because when a synthetic brush is rubbed over leather, it charges the dust in the air and draws it into the bristles. This damages the bristles ... and will damage your shoes. They'll also need replacing sooner than you'd like. A good-quality shoe brush crafted from natural materials will last a lifetime.

Regular cleaning – shoes made from leather are essentially a skin with no body to look after it anymore, so use high-quality creams to hydrate the leather, and wax (non-silicone) to polish them. If you're wearing your shoes a few times a week, a weekly care regime is recommended.

CIRUS STREET SKATEBOARD

Woodworking

A pro skateboarder can go through more than 60 boards a year; a dedicated rookie won't be far behind that figure. Skateboards have to bear the weight of an adult human, are repeatedly slammed against the ground, and are tilted in every direction – all in the name of air, pop and tricks. Most skateboards are made from Canadian Maple, but this one, which is produced in a one-man workshop in central Europe, is crafted from layers of sustainably sourced bamboo and carbon fibre. Why bamboo? Its tensile strength is up to 15 % greater than maple, and once treated, it can be made into uniformly high-quality veneers that provide superb performance. Dániel Bolvári glues these together to make superior decks offering better-than-maple pop, greater resistance to snapping, and a longer life. His upstart brand's catchphrase is 'You break first'. Strangely reassuring.

MEASUREMENTS	Length: 81 cm (32 in); thickness: 11 mm (⅖ in); weight: 1.2 kg (2 lb 10 oz)
MATERIALS	Bamboo veneer, carbon fibre, epoxy resin, nitro varnish
KEY TOOLS	Gluing roll, logo burner, sandpaper, scalpel, screwdriver, varnish sprayer
KEY MACHINES	Eight-headed drill, hydraulic press, heat transfer machine, milling machine, sanding machine, sawing machine
TIME TO MAKE	3 days
LIFESPAN	1 month +

DÁNIEL BOLVÁRI — Skateboard maker [Veszprém-Kádárta, Hungary]

When Dániel, or Dani, first got the keys to the one-time watermill warehouse he'd rented as a maker space for his skateboard company, Cirus Skateboards, back in 2012, it was autumn. Temperatures were around 10°C (50°F), and there wasn't any heating or water, just electricity. But that didn't stop him from getting down to work.

'My workshop is in an old 170 sq m [1830 sq ft] warehouse, next to a small spring. When I first rented it, I separated out a 30 sq m [323 sq ft] space and started making boards. The conditions were harsh; I had to set up a small stove with a narrow iron chimney that wasn't working properly; when the north wind blew, smoke came back in and filled the room! I had to wear several layers of clothes.

'In the past few years I've developed the entire space into a fully functional carpentry workshop. And I have heating and water!'

He also has machines; lots of them, including presses that he made himself. The presses literally press the carefully selected pre-cut layers of bamboo veneer (there are seven in a Cirus Street), epoxy resin and carbon fibre together to make a board.

'After pressing, the raw pressed laminate has to rest for one or two days. It needs to cool down, I let it breathe and stretch. I drill the truck holes then I cut it to its final form; at first roughly with a sawing machine, then more precisely with a milling machine.'

Next, Dani smooths the board by sanding it all over, first with a machine and then by hand. He burns the Cirus Skateboards logo onto the top layer with a simple hand burner that works in the same way as a soldering iron. The board is varnished with nitro varnish then left to dry for another day. Lastly, he uses a heat transfer machine to press on any graphics.

'I love every part of the production process. With each step, I'm getting closer to the final product. The pressing is probably my favourite part; it's the most crucial step. It's when all the different materials come together and when the final form is born. It's also

CARING FOR YOUR SKATEBOARD

This board's striking bamboo surface begs to be left paint-, sticker- and graphic-free. Instead, at the end of each day's skating, sand your board with fine sandpaper (120 grit or so). Pay more attention to the edges than the surface of the board, as that's where any weaknesses will start – and grow, if they're left unattended. After sanding, apply linseed oil with a soft cloth. This will remove any cracks or chips.

> "
>
> I like my tools to be good quality, my blades to be as sharp as possible and everything to be close at hand. If one of my tools is missing, I can feel it.

the riskiest part; if I haven't prepared the glue properly or if something else isn't perfect, the board fails. I'll never get sick of waiting to see how a pressing turns out.'

Dani started skateboarding when he was 15 years old, and first thought of making boards at 17, feeling held back by the price of boards in Hungary (the equivalent of a month's salary) and confident he could make boards to rival the imports he and his friends coveted. He got the chance in 2007, aged 29, when his older brother, Ádám, got involved.

These days Dani still skates, but thinks of himself as a skateboard maker, rather than a skateboarder. He's also a committed environmentalist, and is as pleased with bamboo's environmental superiority as he is with its durability.

'Canadian maple trees take at least 50 years to grow, and skateboard manufacture is the major cause of maple deforestation in Canada. On the other hand, bamboo plants can grow to full maturity in just five years. They improve the land they grow in by preventing soil erosion. Bamboo also traps four times as much carbon dioxide as maple, and emits more oxygen. It's better for skaters, and better for the environment.'

In the past, Dani's had jobs he didn't want to go to. But in the workshop, surrounded by his tools and boards, it's different.

'Sometimes I'm desperate to get into my workshop. Many times I've spent whole weekends here, just to be working and mucking around. I don't have a specifically favourite tool, they're all important. I like my tools to be good quality, my blades to be as sharp as possible and everything to be close at hand. If one of my tools is missing, I can feel it.'

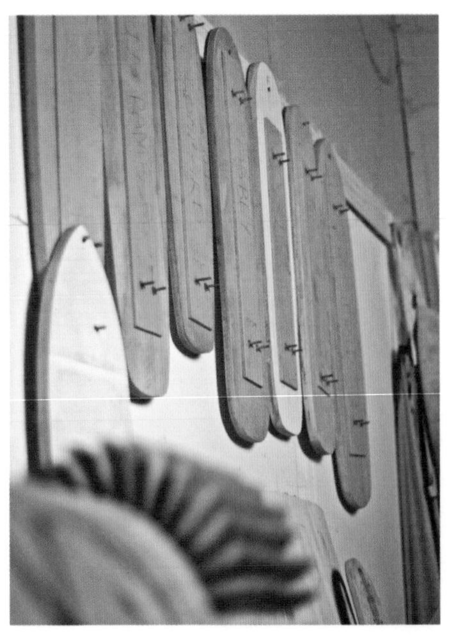

HOW TO

INSTALL TRUCKS AND WHEELS ON YOUR BOARD

MATERIALS

- Deck
- 2 trucks (each should come with: 4 truck screws, 4 corresponding nuts, 4 axle washers, 2 locknuts)
- 8 wheel bearings
- 4 matching skateboard wheels

TOOLS

- Adjustable spanner (wrench)
- Phillips screwdriver or all-purpose skate tool

METHOD

1. Put the eight truck screws through the holes on the top of the deck, then flip the board over so it's bottom side up.
2. Slip the trucks onto the screws, through the baseplate. Position each truck so that the kingpin (the big pin at the centre of the truck) faces inwards, towards the opposite truck.
3. Loosely attach the nuts to the screws to hold the trucks in place until it's time to fasten them securely.
4. Use your adjustable spanner (wrench) to hold the nuts in place, then use the screwdriver or skate tool to tighten the screws in a crisscross pattern: start by tightening the north-west screw, then the south-east one, then the south-west, then the north-east. Make them tight, but not too tight, lest you crack your deck. This will ensure that the hold is evenly distributed over the entire baseplate of the truck.
5. Secure the other truck, then it's time to add the bearings and wheels.
6. Flip your skateboard onto its side and slide a bearing onto one of the truck's axles. Take care here – bearings are delicate and their integrity will be compromised if you don't handle them carefully. Follow the bearing with a wheel.
7. Push down from above to force the bearing into the wheel socket. You'll need to push pretty hard; eventually, you'll feel the bearing slip into place.
8. Once the first bearing is in, remove the wheel and slide on another bearing. Flip the wheel over, slide it onto the axle, and push it into place.
9. Repeat this process with the other three wheels.
10. Once you've installed two bearings on each of the four wheels, it's time to attach the wheels to the trucks. Begin by setting the board on its side again.
11. Slide on a washer, then the wheel with bearings, and then the second washer. The last piece to go on is the lock nut. Tighten this with the spanner (wrench) to secure the entire wheel – again, tight but not so tight as to restrict the wheel's motion. Spin it to see how it moves.
12. Attach the other wheel on that side of the board, then flip the board and repeat the process on the other side.
13. Get your skate on.

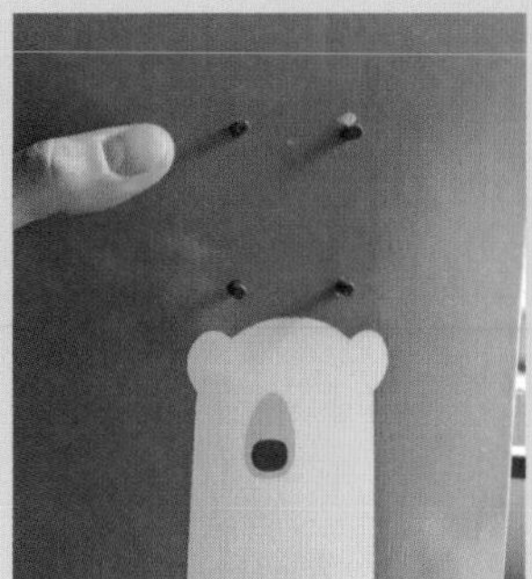

Step 1

Step 11

WWF
CIRUS
GIOTTO
ROBERCOLOR
NOSE

VALPARAISO - CHILE
SKATEDELUXE

CRAFTSMAN
SOAP COMPANY
BEARD
SOAP
CRAFTSMANSOAP.COM
NET WT. 4 OZ

BEARD SOAP

Soap making

A beard can be worn many ways. Long and lean, short and sturdy, tidily trimmed or as wild as Karl Marx crossed with a yeti. However you like to wear your facial hair, a decent beard care regime will help keep it within the bounds of social acceptability. And that's where beard soap comes in. Hand made from a carefully selected blend of base oils, butters and essential oils by Evan Worthington of Los Angeles' Craftsman Soap Company, this earthy smelling bar is formulated to build a particularly thick, creamy lather for a thorough beard clean that won't leave your beard or face dry. Use it to build a lather in your hands or directly against your beard, working it like a shampoo from the roots of your beard to the tip. A bearded chap himself, Evan is committed to helping make beards last longer, one bar of soap at a time.

MEASUREMENTS	Weight: 114 g (4 oz); size: 8.5 × 6 × 2.5 cm (3½ × 2½ × 1 in)
INGREDIENTS	Castor oil, cedarwood essential oil, cocoa butter, coconut oil, lavender essential oil, lye, olive oil, rosemary leaf extract, shea butter, spruce essential oil, tea tree essential oil
KEY TOOLS	Card scraper, droppers, scales, measuring cups, mixing wand, ratchet straps, wood and wire frame cutter, wood moulds
KEY MACHINES	None
TIME TO MAKE	24 hours to mix, set, and cut; 4–6 weeks to cure and dry
LIFESPAN	3 years

EVAN WORTHINGTON — Soap maker [Los Angeles, USA]

Evan Worthington spent years making bars of soap for his own use and as gifts for friends and family before some encouragement from his sister inspired him to try his hand as an entrepreneur. Back then, in 2012, he worked on his kitchen stovetop; today he has a bustling 93 sq m (1000 sq ft) workshop on the outskirts of Los Angeles.

'Growing up I was always trying to understand how things worked. Everything from carpentry and metalwork to electronics and circuitry has captured my attention at some point; with soap making I realised I could scavenge goods from grocery store aisles and make something entirely new from basic ingredients. It's this really simple, elegant acid-base reaction where the sum of the parts is this natural, aromatic, utilitarian product. It's a high-school science experiment that I continue to find endearingly magical,' says Evan.

Evan works in small batches, making blocks of soap 'the size of half a case of beer' that he cuts down to size with a guitar string. But of course, first, he must create the recipe.

'For the beard soap, I tried to strike a balance between essential oils that build an appealing fragrance, and ones that have practical benefits for hair care. Cedarwood and lavender essential oils can stimulate circulation of hair follicles, and tea tree essential oil moisturises.'

The beard soap only took a few goes to get right. Others in Evan's range, like the shaving soap, took more than 20 iterations. Once the recipe is ready to go, there are five steps to soap making – measuring, mixing, cutting, curing and drying.

'Measuring for soap making has to be precise because you have to balance the low pH of the oils with the high pH of the lye to create soap properly. For consistency, I do all my measuring by weight on a gram scale. I use slightly less lye than is necessary to convert all of the oils to soap, as this leaves small amounts of moisturising and conditioning oils and essential oils in every bar.'

Evan blends everything together – the oils, melted butters, essential oils, clays and botanicals – then adds a prehydrated liquid lye solution. Lye is a highly soluble liquid metal hydroxide that aids saponification, helping to dissolve, blend, heat and thicken the ingredients.

'I always wear goggles and gloves when handling lye, as it can burn the skin. I stir the mixture vigorously with a drill-driven mixing wand to make it react with the fatty acids in the oils; the lye is completely consumed in this process. As the reaction occurs the liquid thickens and I pour it into the wooden moulds, which I make myself.'

The next day, the previously liquid soap is a firm but pliable block. Evan unmoulds it and cuts it into chunks with a guitar string. He leaves the chunks to dry for a few hours to form a firm outer later that is similar to a cheese rind, which makes the soap easier to handle when cutting it into individual bars.

'Next I stack the bars on a drying rack, making sure they're evenly spaced. They cure over the next few days, which is the tail end of the soap making process, where any residual saponification occurs. What is left in the soap is technically a blend of soap salts (reacted lye and fatty acids), residual pure oils and butters, and any clays and botanicals added to the soap. I leave the fully cured soap on the racks for several weeks to dry and harden, which makes the soap last longer when it's put to use.'

Evan's handcrafted processes make each bar truly unique, from a live edge that reflects the shape of the mould to small marks of character and the impressions from the cutting process.

'Being in the workshop and making soap is one of those physical tasks that helps me find total focus; time slips away. Of course, there's also an aroma-therapeutic element – splashing around various essential oils is definitely uplifting.'

> “
>
> ... with soap making I realised I could scavenge goods from grocery store aisles and make something entirely new from basic ingredients ... It's a high-school science experiment that I continue to find endearingly magical.

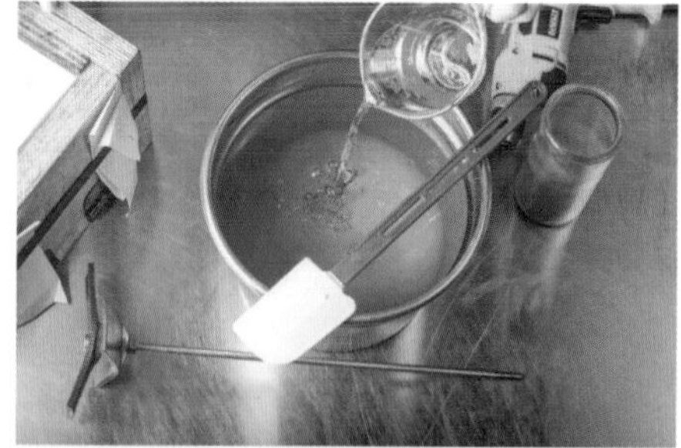

CARING FOR YOUR BEARD SOAP

Keeping natural soap as dry as possible is the best way to make it last. Keeping it out of the shower stream and in a draining rack or dish will make a big difference, otherwise cut off small slices to use as you need them, rather than using the whole thing. Putting it in the sun or moving it to another (drier) room for a day or two can also extend its lifespan.

HOW TO

MAKE YOUR OWN SIGNATURE BEARD OIL

Evan makes beard oil as part of his Craftsman Soap Company range too. This isn't a recipe from one of the oils he makes for his label, but a simpler method he has developed for home use.

INGREDIENTS

- Grapeseed or almond oil (light-bodied oil)
- Jojoba (heavy-bodied oil)
- Coconut (heavy-bodied oil)
- Lavender, sage and eucalyptus essential oils (or other oils you find appealing)

TOOLS

- Scale (optional)
- Pyrex measuring cup
- Microwave oven
- Two empty bottles
- Eyedropper (optional)

PRO TIPS

- *Coconut oil melts just above room temperature and can be melted quickly at the lowest temperature on your oven, or even on the warm surface of an oven in operation (between ranges, not on a burner). During winter you may need to warm it in the microwave, but on a warm day just an hour or two in a sunny window will do the trick.*
- *Undiluted essential oils should never be used directly on the skin.*

METHOD

1. Think about what kind of oil you and your beard might prefer. Light-bodied oils spread easily and penetrate quickly. They leave little sheen and dry quickly. Heavy oil blends can still penetrate, but will leave behind a finish, giving your beard some sheen and a bit of weight that can help shape the beard. Jojoba is actually a liquid wax, which helps coat and protect, but may be heavier than you prefer.
2. Draft the recipe for your base. For the lightest oil, just use grapeseed oil (almond oil is a good substitute if you can't find grapeseed oil), or start with a heavier ratio of 2:1:1 using grapeseed, jojoba and coconut oils.
3. Keep it small; work with increments you can accurately measure, but leave plenty of room to tweak the recipe to your preferred weight and proportions.
4. Measure and blend. Use a scale to achieve exacting proportions, or work by volume in the measuring cup if you don't have a scale. If your oils have hardened, gently melt them in a microwave or by placing them in a sunny spot.
5. Now draft the recipe for your essential oil blend. Evan suggests lavender, sage and eucalyptus, as they are common and affordable oils, but use what you like. If you're sampling and there are testers, put two testers side by side and smell them together to get a rough idea of how those particular oils will meld. A simple 2:2:1 ratio of lavender, eucalyptus and sage is a good place to start, and counting drops is the most accurate method of measuring (unless you have a gram scale).
6. Ideally, blend up about 60–100 drops of oil, then bottle it separately to your base in a smaller bottle. If you have the patience, let it rest – at least overnight and up to a week or more – to meld. With time, even discordant essential oils can become surprisingly harmonious.
7. Blend your base and fragrance together by adding the fragrance to the base. For accuracy, use an eyedropper. Strength is subjective, but 20–30 drops per 30 mL (1 fl oz) of base oil is a good starting point.
8. To use, apply oil after bathing, when your beard is still damp, starting with just a few drops in your palm. The water in your beard will help disperse the oil, allowing you to get the most out of your beard oil. Spread it evenly down to the roots with your fingertips, then draw it out to the tips by hand or with a brush or comb.

Step 4

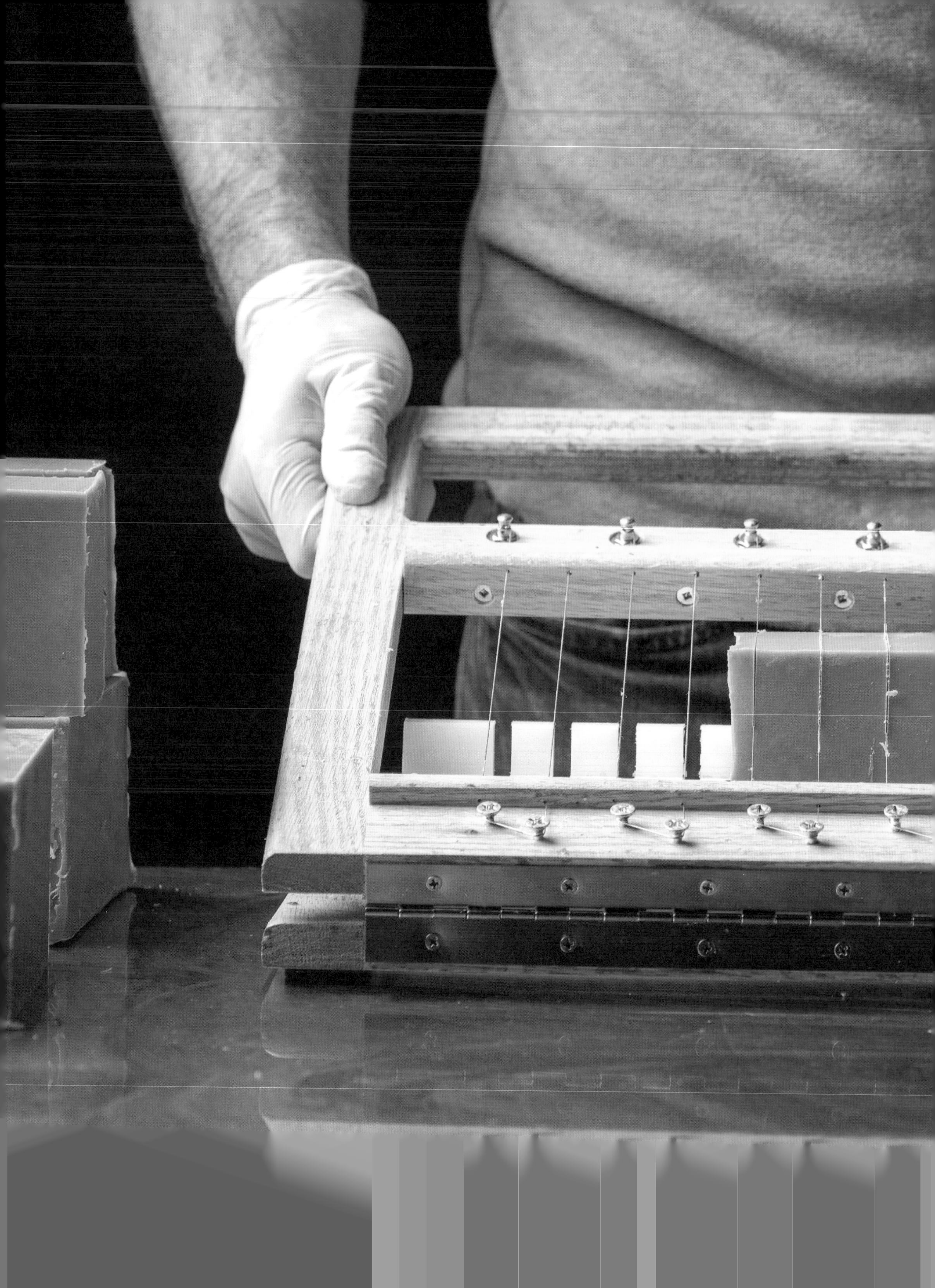

WOOD GRAIN BOOMCASE SOUND SYSTEM

Electronics

Meet the BoomCase, an up-to-the-minute, self-powered digital sound system housed in a repurposed vintage suitcase that's been gutted and reinforced, and had speakers and a battery pack built into its frame. Designed and made by a pair of enterprising Californian brothers and experimental sound makers, Dominic and JP Odbert, the BoomCase can receive sonic input from any music playing device that has a headphone output. It has optional Bluetooth connectivity and USB facility, charges via a laptop-style cable and plug, and can pump sound for 18 hours on a single charge. Each of BoomCase's portable sound systems is a one-off production. The Odbert brothers will even build a speaker system into a suitcase of your own supply. Your Great Uncle Edward's portmanteau, perhaps? Now what would the old boy make of that?

MEASUREMENTS	Length: 53.3 cm (21 in); height: 35.6 cm (14 in); depth: 17.8 cm (7 in); Weight: 9.1 kg (20 lb)
MATERIALS	8 in woofers, copper wire, charger port, custom built rechargeable battery pack, digital 400-watt amplifier, dome tweeters, mid-range speaker, power switch, vintage Samsonite suitcase, wood stain and finish
KEY TOOLS	Phillips screwdriver, soldering iron, wire stripper with crimper
KEY MACHINES	Hand drill, handheld router, orbital sander
TIME TO MAKE	7 days
LIFESPAN	10 years +

DOMINIC ODBERT — Portable sound system maker [California, USA]

Sick of continually having to buy D-sized batteries in batches of 12 for his BoomBox (which didn't even sound that great), in 2009, Dominic Odbert decided to build his own portable sound system. He'd always loved the look and feel of 1950s record players, so the inspirational leap to upcycling vintage suitcases for his creations was a quick one.

Dominic got started on a prototype, along with his older brother, JP. The two brothers are no strangers to portable sound. For several years, a 1950s army truck that they decked out with 18 subwoofers, eight full range speakers and a dance platform known as the Temple of Boom was a staple in the annual LovEvolution Parade, which techno-fied the streets of San Francisco from 2004 to 2009.

'We've been making experimental electronic music and building synthesisers together since 1999, but we wanted to come up with a way to carry our music with us that was more portable than a truck. But for music to be portable, its casing needs to be tough. I wanted the BoomCase to have a built-in rechargeable battery, and to sound great,' says Dominic.

Dominic sources his vintage suitcases, the occasional lunchbox and briefcase, old radios, and vintage barrels from antique stores and estate sales.

'They're so sturdy; they were made well over 50 years ago and they're still in amazing shape. Many are even older, dating to the thirties and forties. Some have tags on them from actual people; we like to leave those on the cases so that they travel forward with a bit of their history intact.'

Construction of a single BoomCase takes about a week. Every new build starts with a design process. 'Every suitcase we work with is unique. We like to create a unique speaker design for each BoomCase too. We design with the acoustics in mind; how can we get the best sound out of the speaker and suitcase once they're put together?'

This particular model, the Wood Grain BoomCase, was made by stripping a regular vintage suitcase down to its bare pine frame and finishing the wood. Dominic prepares the wooden surface by sanding it down, first with an orbital sander and then by hand, then laying out and marking up the speaker design on the actual suitcase.

Next, he uses a hand-operated router and templates to cut out the holes for the speakers, then stains and finishes the suitcase by applying a wood stain with a cloth.

'To make the suitcase even stronger than it is, we build another box within the suitcase to produce better overall acoustics.'

Now Dominic prepares the suitcase to take the speakers – for this one, that's two 8-inch woofers, two dome tweeters and a mid-range speaker. He uses an electric hand drill to drill holes for the switch, charger port, audio input and volume control, then installs them. He installs the speakers and reinforces the installation by sealing the suitcase so it's airtight. Next come the amplifier and battery, then he wires up all the components. The most important tool here is an all-in-one wire stripper and crimper, which enables him to easily strip wires to the lengths he needs, and crimp them.

Dominic seals the BoomCase using wood stripping and a gasket foam around the edges, then spends a couple of days testing it.

'I test various types of music at medium to maximum volume to make sure everything's working to audiophile sound quality standards before sending one out.

'The most challenging part of each BoomCase build is engineering the best possible crossover for each system. A crossover is what makes each speaker play only a certain range of frequencies. Getting this just right is the key to our sound quality.'

CARING FOR YOUR BOOMCASE

The BoomCase is housed in a hardy suitcase that may have quite literally travelled the world; there isn't much you can do that will mess it up. To keep the battery in good shape, make sure you charge it at least once every six months. If, after a few years, the wood surface seems tired, a quick, light wipe down with some tung oil and a clean cloth will bring the sheen back and protect it for many more years to come.

“

Every suitcase we work with is unique ... We design with the acoustics in mind; how can we get the best sound out of the speaker and suitcase once they're put together?

HOW

AUDIO CROSSOVERS WORK

Audio crossovers are a type of electronic filter circuitry that are used to split an audio signal into two or more frequency ranges in a range of audio applications. People can hear sound frequencies from 20–20,000 Hz – but there's just no way a single speaker can handle such wide-ranging frequencies without seriously mangling the sound.

Enter crossovers. They enable the signals to be sent to 'drivers', or speakers designed for different frequency ranges – tweeters for high frequencies, woofers for midrange frequencies and subwoofers for low frequencies. When well designed and made, crossovers will give your audio ear-pleasing relative volume, and eliminate distortion.

They're often described as 'two-way' or 'three-way', which indicates how many frequency ranges the crossovers split the signal into.

A two-way crossover, also called a parallel crossover, is the simplest and most common type. Parallel crossovers share the same input for each driver. Once it departs the driver, each crossover has its own 'branch' of the circuit, and is essentially independent of the other. Parallel crossovers are suited to audio systems that involve a woofer or mid-woofer and a tweeter. These include basic home and car audio systems.

In a three-way crossover there's the woofer and tweeter, plus a mid-range speaker to provide more depth to the audio. They're usually found in more complicated (or high-end) home audio speakers and car audio systems.

Crossovers are also defined as 'passive' or 'active'. A passive crossover splits up an audio signal after it is amplified by a single power amplifier, so that the amplified signal can be sent to two or more driver types, each of which represents different frequency ranges. These crossovers are made entirely of passive components and circuitry. The term tells us that no additional power source is needed for the circuitry – it just needs to be connected to the power amplifier signal.

Active crossovers require power and ground connections, but give you much more flexibility and the ability to really fine-tune your music. An active crossover contains active components in its filters and is operated at levels suited to power amplifier inputs. Active crossovers demand the use of power amplifiers for each output band – a two-way active crossover needs two amplifiers, one each for the woofer and tweeter.

Dominic builds the crossover from scratch on each BoomCase build. For most BoomCases, he goes with a three-way crossover, as he likes the way a mid-range speaker fills out the audio spectrum.

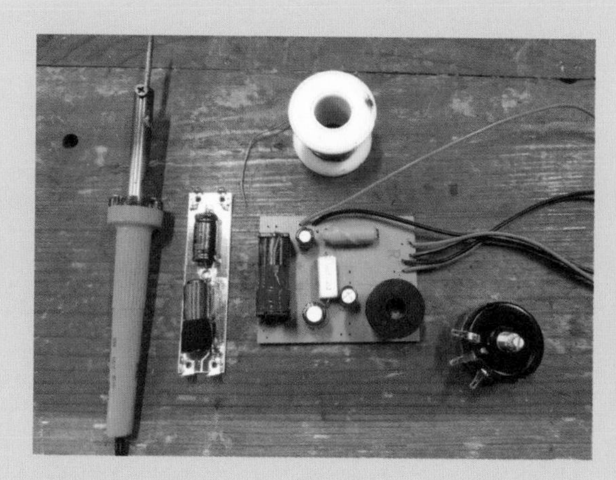

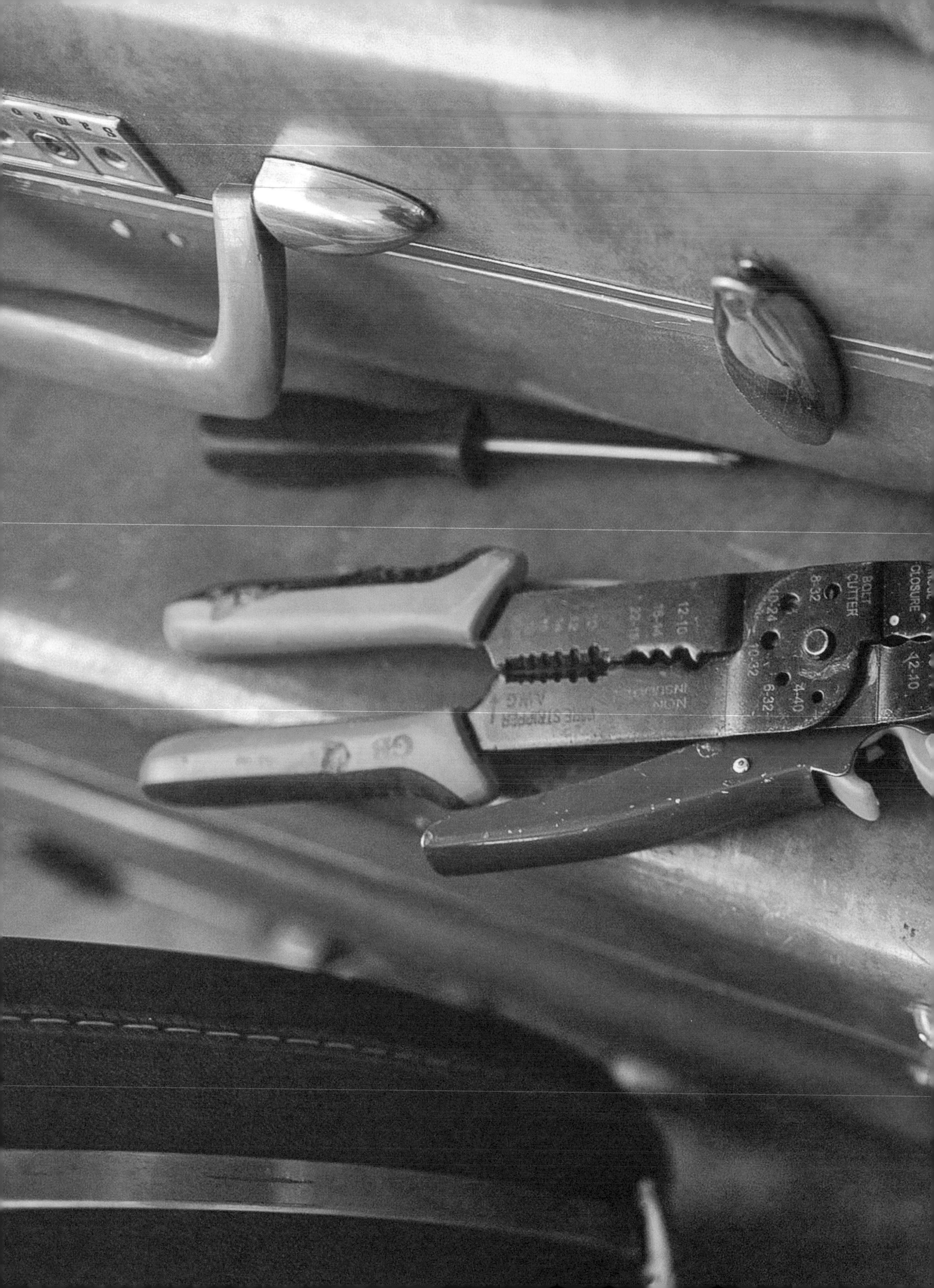
BOLT CUTTER
CLOSURE
8-32
10-24
10-32
6-32
4-40
12-10

RING-NECKED PARAKEET SPECIMEN

Taxidermy

Taxidermy is the art of preserving animal skins, and has been practised since ancient times. The word 'taxidermy' literally means 'the arrangement of skin', and it is a skilled craft that demands sculptural and anatomical knowledge and a love of natural history. It has been deathly popular since the 1700s – even Theodore Roosevelt was a fan. Modern practices focus on making specimens look as lifelike as possible. This ring-necked parakeet was lovingly preserved by Jazmine Miles-Long, an ethical taxidermist in the United Kingdom who only works with animals that have died from natural causes. The parakeet flew into a window and died, and was donated to Jazmine, who is a vegetarian, by the woman whose window it flew into. She contacted Jazmine via Twitter (how fitting!), and the bird now rests eternal in the Western Park Museum in Sheffield.

MEASUREMENTS	Wingspan: 46 cm (18 in); length: 40 cm (15½ in); weight: ~500 g (17½ oz)
MATERIALS	Acrylic eyes, balsa wood, clay, cotton thread, cotton wool, galvanised wire, homemade bird preservative paste, ring-necked parakeet, wood glue
KEY TOOLS	Gloves, hair dryer, knives, metal brush, pliers, scalpel and blades, scales, sewing needle, tweezers of all shapes and sizes, wire cutters, pins
KEY MACHINES	Freezers
TIME TO MAKE	2 weeks
LIFESPAN	100 years +

JAZMINE MILES-LONG — Ethical taxidermist [Hastings, United Kingdom]

In 2007, Jazmine Miles-Long volunteered at the Booth Museum of Natural History in Brighton, which focuses on Victorian-era taxidermy. She spent most of her time renovating taxidermy specimens, but was lucky enough to be shown the basics of taxidermy by the curator. Fascinated, she enrolled to study the art shortly after, and has been providing bespoke taxidermy services and teaching taxidermy ever since.

When a new specimen comes into Jazmine's workshop, it goes straight into the freezer. This is to kill any fleas or ticks on the animal; working on a fresh specimen without freezing it means that Jazmine risks transferring the pests to her own body.

When she is ready to work on the specimen, Jazmine puts on a pair of latex gloves then places the defrosted specimen on her workbench; one of two in a studio she shares with her partner, illustrator and artist Benjamin Phillips, in Hastings, East Sussex.

Jazmine stands at her bench to work. A cabinet filled with glass eyes sits on one corner; anatomical drawings of various animals fill the surrounding walls. 'My tools are all over the place, as I have so many for different tasks. I preserve specimens, make cases and create ceramic flora for the cases. But I always have to hand on my bench a white plastic tool box with tweezers, knives, blades, pliers and gloves inside – the most immediately needed bits,' says Jazmine.

She begins by weighing and measuring the specimen. Then comes skinning. Jazmine works with her tweezers and scalpel, carefully slicing and pulling. The first incision on the parakeet was along the breastbone, down the centre of the body.

'People often presume this will be very messy, but it's not that bad. Underneath the skin is a membrane that acts like a second skin. It keeps the body together in one piece, and working carefully between these layers means the skin can be simply peeled away. On a mammal the skin must be pickled and tanned, a bit like leather. But with birds, all of the fat must be cleaned away from the feather tracts where the quills poke through on the inside of the skin.'

After this has been done, Jazmine washes the skin and dries the feathers, taking care to keep the inside of the skin wet. Next, she crafts the form that will go inside the skin to replace the muscular structure.

The parakeet's interior was carved from balsa wood, using a hacksaw, Stanley knife, and sandpaper. Jazmine retained the bird's skull, cleaned it using tweezers and cotton wool, keeping it attached to the skin all the while. She also kept some of the wing and leg bones attached to the bird's skin, with all the flesh cleaned away, to help maintain the shape in those areas. Wire and cotton thread were used to make the neck, wings, legs and tail. The eye sockets were filled with clay, the acrylic eyes were carefully put into place, and the skin pulled back over the skull. The components were all carefully arranged inside the parakeet. Then Jazmine brushed her homemade preservative paste onto the inside of the parakeet's skin and sewed it up.

This is the bulk of the work, and on a bird like the parakeet, takes around 2 days to complete. Jazmine pins the specimen into place, then leaves it to dry for about a week. It's then unpinned, ready for the final stage.

'Once a piece has dried, any skin not covered by fur or feathers loses its colour and turns a dark yellow or grey. This is common around the eyes, inside the ears, on the pads of the feet in mammals and on the legs and bills of birds. The last thing I do to preserve the specimen is to paint these areas using acrylic paints so they look natural.'

Finally, she fits a specimen onto a base or into its case, ready for display.

'Taxidermy helps people engage with the natural world. It gives us an opportunity to get up close to animals that we would struggle to ever see or approach in the wild; I feel honoured to be able to work with animals in this way.'

> Taxidermy helps people engage with the natural world. It gives us an opportunity to get up close to animals that we would struggle to ever see or approach in the wild; I feel honoured to be able to work with animals in this way.

CARING FOR YOUR SPECIMEN

A modern taxidermy specimen should last a very long time; some museum pieces date back to the late 1700s. However, current restrictions and practices mean 21st century taxidermists do not use poisons to preserve specimens (and ward off attackers) as their predecessors often did. This means that modern taxidermy is at risk of insect attacks – think moths, dust mites and carpet beetles.

If you do see evidence of moths or dust mites on your taxidermy piece, place the mount into a sealed plastic bag and store it in the freezer for at least two weeks. This will kill any infestation.

Specimens are best kept dust free in a sealed case in a climate controlled area, away from dampness and out of direct sunlight. Do not use any cleaning product on feathers or fur, simply dust your mount using a feather duster. Touching your taxidermy is very tempting, but oils in the skin will eventually damage the feathers and fur, so try not to handle your specimen unless necessary – and if you do, wear latex gloves.

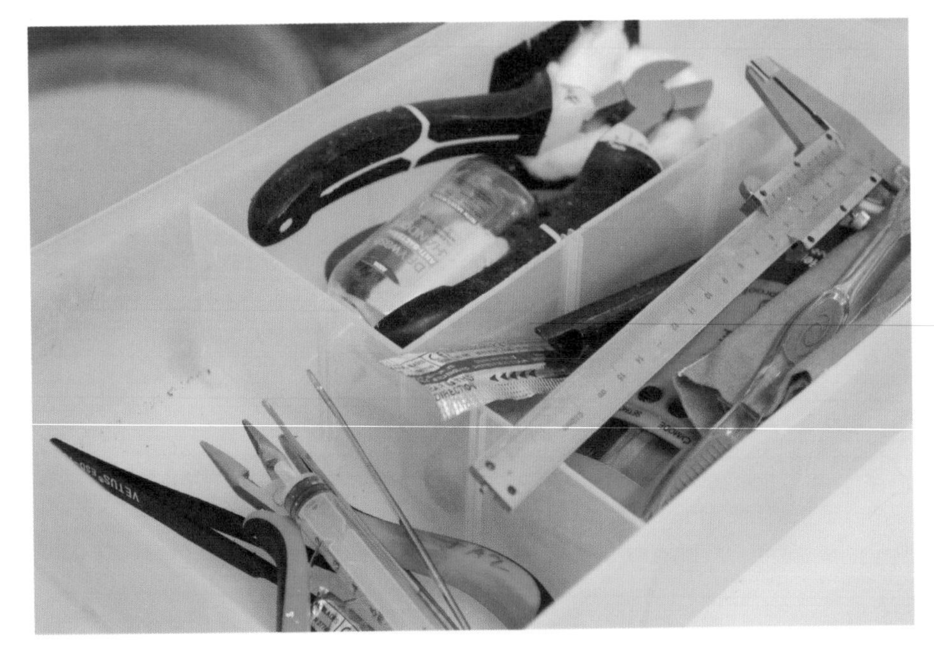

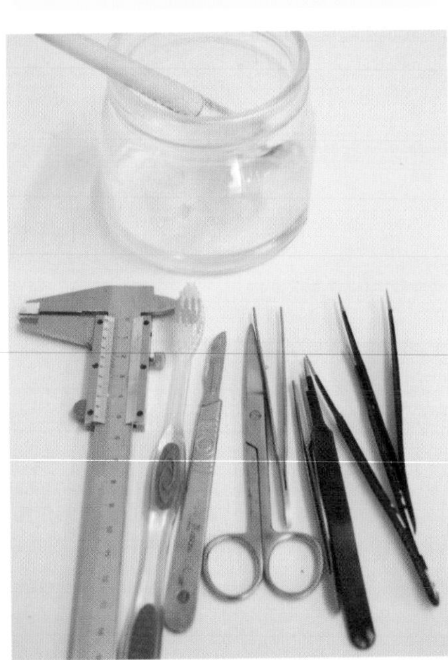

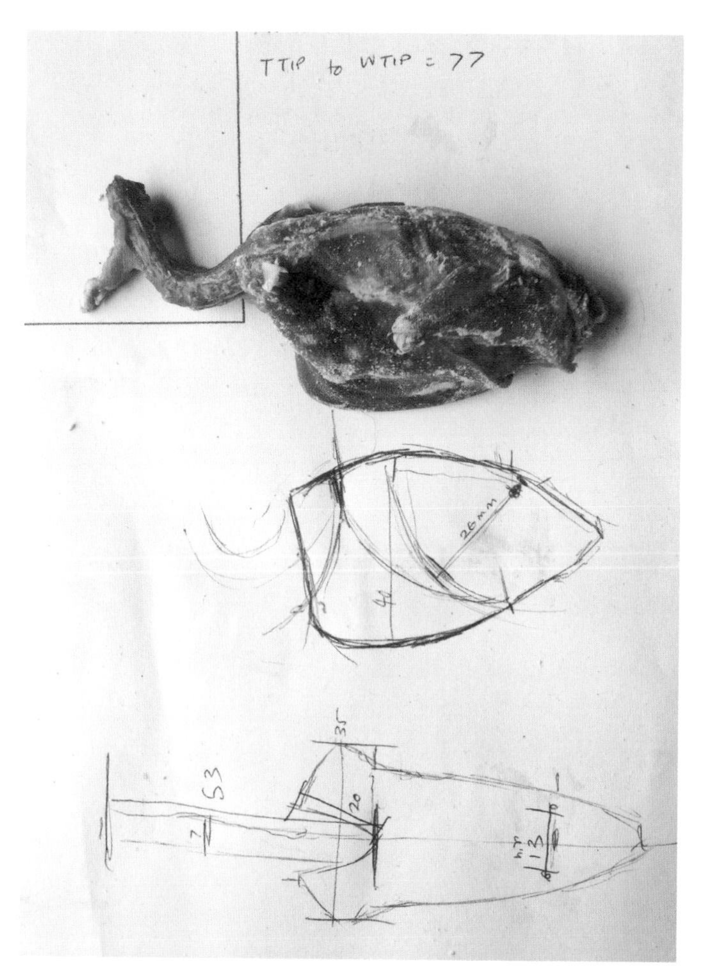
TTIP to WTIP = 77
26mm
35
53
20
13

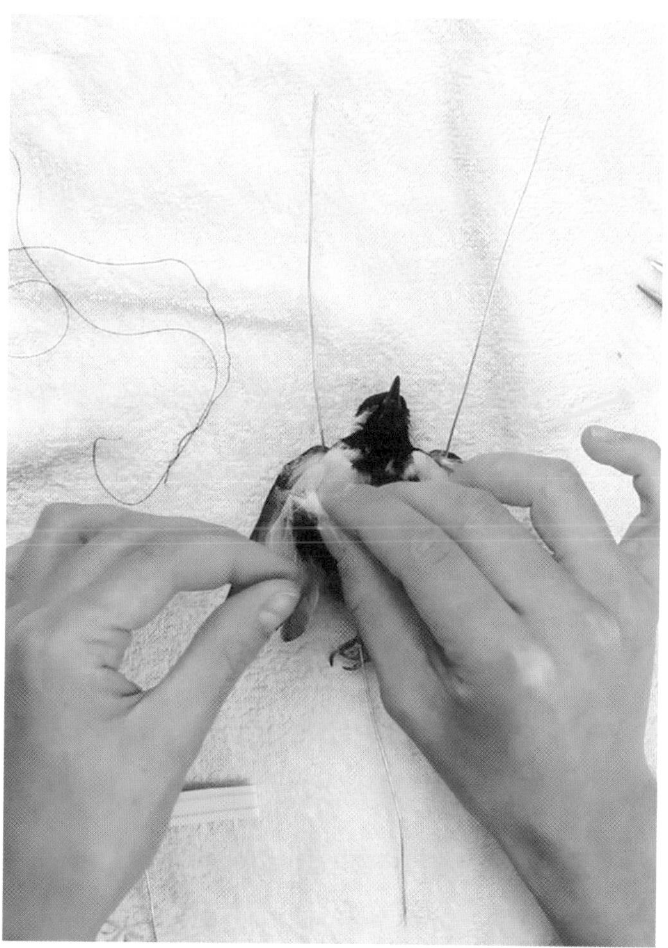

Birds of East Asia
BIRDS
MAMMALS
BOOTH MUSEUM OF NATURAL HISTORY
BRIGHTON

HOW TO

PRESERVE A BIRD'S WING

Taxidermists will often preserve a bird's wing (rather than the entire bird) when a bird is too damaged to use for full taxidermy. This is an excellent way for interested novices to test out their appetite for taxidermy.

MATERIALS

- Bird (died from natural causes)
- Water (cold, in a jar)
- Washing-up liquid
- Salt
- Cotton thread
- Piece of cardboard (larger than wing)
- Card

TOOLS

- Paintbrush (any)
- Scalpel
- Tweezers
- Basins (one filled with cold water and another with water, salt and washing-up liquid)
- Towel
- Sewing needle
- Hair dryer
- Toothbrush
- Pins

SAFETY GEAR

- Latex gloves

METHOD

1. Always write down the species of bird, whether it was wild or captive bred, its sex, age, where you found it, how it died and the date it died, and your own name and address. These details can be written onto a small label that can be put with the finished wing for your records, proving where the bird is from and that it was legally obtained. Laws around the world vary, and Jazmine advises that you should always check that the species you are working on does not have any specific restrictions or paperwork needed in your country.
2. Freeze the specimen overnight to kill off any fleas or ticks, then defrost it before starting work.
3. Put on the gloves and place the defrosted specimen on a paper towel on your intended work area.
4. Take the paintbrush and dip it in the jar of cold water.
5. Gently take the feathers around the shoulder of the bird, wet them with the paintbrush, then part them to expose the skin.
6. Use the scalpel to cut through the skin and remove the wing at the shoulder joint, making sure you guide the blade in between the joint, and not through the bone.
7. Once the wing has been removed, wet and part the feathers along the bone on the underside of the wing and use the scalpel to make two incisions along the bones on either side of the wrist joint.
8. Open out the skin using the tweezers.
9. Use the tweezers and the scalpel to remove all of the meat from the bones. Do not remove the quill ends of the feathers from the bone, or the wing will fall apart. The feathers need to stay attached to the bone for it to function like a wing.
10. Once all the meat has been removed, wash the wing in a basin of cold clean water with a sprinkle of salt and a good dose of washing-up liquid.
11. Do this one more time in a fresh batch of water, then wash it in one or two more basins of cold water, using no washing-up liquid, until the water is clear.
12. Squeeze out as much water as you can and then lay the wing flat on a fresh towel, using the towel to dry off some of the moisture.
13. Thread the cotton onto the needle, then sew up the incision on the inside of the wing. Use a hair dryer and toothbrush to dry the feathers, always brushing with (rather than against) the feathers' natural direction.
14. Position the wing on the cardboard and use the pins to keep it in place.
15. Cut strips of card and pin these down on top to make sure the feathers stay in the desired position.
16. Leave the feather to dry for one week in a dry, well-ventilated room.
17. After it has dried, unpin the feather from the card. It's now ready for framing and display. If you keep it away from direct sunlight, dust and moths, it will last a very long time.

Step 9

Step 10

Step 11

Step 13

Step 13

Step 13

Step 14

Step 15

SYLKO
EYES
GOLD
SIZE

DEWALT

SALVAGE STOOL

Furniture making

The stool is one of the earliest forms of seating. Artefacts of three-legged stools from the 17th century still exist, but it's likely that stools have been around much longer than that. In Africa, for example, they have long been used as thrones. This one, the Salvage Stool, is crafted in New Zealand's capital city, Wellington, from recycled cast aluminium and wood rescued from skips, wood turners' patterns, building recyclers and tip shops. Fixed together with three stainless steel bolts and easily dismantled for transportation, repair and end of life recycling, the Salvage Stool is an exercise in lightweight knockdown design. It's a cheeky and charismatic piece of furniture, too. It'll stand up to being knocked about and will look at home pretty much anywhere, thanks to its sturdy juxtaposition of a refined industrial aesthetic and junkshop outcasts.

MEASUREMENTS	Depth: 33 cm (13 in); width: 33 cm (13 in); height: 42 cm (16½ in)
MATERIALS	Cast aluminium (recycled), balustrades or wooden dowel, recycled wooden legs, stainless steel bolts
KEY TOOLS	Hand files, Japanese saw, skew chisel, square chisel, gouge chisel
KEY MACHINES	Angle grinder, drill press, drop saw, power file, shot blaster, wood lathe
TIME TO MAKE	1.5 days
LIFESPAN	Heirloom quality

TIM WIGMORE — Furniture designer [Wellington, New Zealand]

When Tim Wigmore needs a break from his work, he plays ping pong. There's a table set up in the lounge area of the 230 sq m (2475 sq ft) studio he shares with a luthier and another furniture maker. Competition is fierce, with singles and doubles, intra and inter crew and personal vendetta matches all on the cards.

'The space used to be a surfboard factory. It's by the beach in Lyall Bay, with lots of natural light and high ceilings. In our office, there's a beautiful build-up of coloured resins on the floor from its years as the glassing bay for the surfboards,' says Tim.

The Salvage Stool is one of the first pieces of furniture Tim designed under the brand Designtree, which he launched with his partner, Rebecca Asquith, in 2011. The stool's design and production combine modern technology with traditional craft techniques: 3D printing and aluminium sand casting, with salvaged materials.

Tim designed the original form for the stool top in CAD software and 3D printed it in a sandstone material.

'The shape is repeated three times for each seat top, however only one master is 3D printed, as the sand casting process allows for repetition of the same shape,' explains Tim.

'We give this 3D printed master pattern to our local foundry, who pack it in special sand in a timber frame in three parts, then carefully remove the original to leave a negative form in the sand mould. This process is repeated using the single master pattern until the desired number of moulds have been made. Each mould has holes for pouring molten metal into, and smaller holes which release air and gases to ensure that the metal reaches all parts of the mould.'

After the foundry workers have poured the molten aluminium into the sand, they leave it overnight to cool. Then the sand is pulled away and broken off the cast metal to reveal the positive form inside. The foundry then sets about 'chasing' (cleaning up) each seat's three aluminium parts with hand files, a grinder, die grinders and a power file. Then they're returned to Tim.

'I give the parts an even, textured finish by polishing them with steel beads in a shot blaster, then lacquer them with a clear matt lacquer to help prevent oxidisation of the aluminium and stop the metal absorbing oils.'

Then he starts working on the legs. 'I turn down the end section of the timber leg on the wood lathe using a skew and a square chisel, and sometimes a gouge. I assemble the stool in order to mark the leg lengths and drill the bolt holes. From here I dismantle the stool and cut the legs level with the floor and flush with the top of the stool using a combination of a Japanese saw and a drop saw.

'I place the legs in a jig in the drill press and drill holes for the stainless steel bolts (one per leg). I hand sand the cut ends of the legs with a block and paper. At this stage I re-check the fit of the parts and make adjustments as necessary, then I use a natural citrus-based resin oil to seal the cut surfaces and faces if required.'

The stool is flat packable, so once the legs are dry, Tim packs the top, legs, bolts and care instructions in one of Designtree's custom boxes, labels it, and it's ready for shipping.

'Most of the legs are sourced from wood turning shops; they're often old balustrades and chair legs that didn't get used. But I'll consider anything for legs; I've used tool handles, dowel offcuts, 2 × 4, even tree branches. That's the premise of the Salvage Stool, it takes unwanted objects, gives them a new use, and makes us see them in a new context.'

CARING FOR YOUR SALVAGE STOOL

Oil the wooden legs of the stool with furniture oil every couple of years to nourish and protect the timber. Keeping the stool out of direct sunlight and away from excessively wet or dry areas will prolong its life. Ensure the aluminium is kept dry. It is a robust material, but if the surface becomes badly worn or severely scratched, it should be re-blasted and re-sealed to prevent oxidisation.

“ I'll consider anything for legs; I've used tool handles, dowel offcuts, 2 × 4, even tree branches. That's the premise of the Salvage Stool, it takes unwanted objects, gives them a new use, and makes us see them in a new context.

HOW TO

DESIGN FOR 3D PRINTING

Tim's furniture making combines traditional and contemporary design and manufacturing techniques. He talks us through his approach to design for 3D printing.

TOOLS

- Pen or pencil
- Paper
- Scanner (optional)
- Design software with 3D capability, such as CAD, Rhino, Solidworks, TinkerCad, FreeCad or Sketchup

PRO TIP

- *If you don't have to have a 3D printer of your own, outsource your printing to an online 3D printing service and have the model posted out to you. This will likely save on cost, and give you greater options when it comes to materials and internal lattice work.*

METHOD

1. Make a pen or pencil sketch on paper of the form you'd like to create. Try to figure out as many details as possible – think scale, proportions and angles – before getting started on the computer, as it's quicker and easier to get details like these sorted on paper.
2. Once you are happy with your sketch, scan it and import it into your CAD software. Tim finds that using his pencil sketch as a base for tracing over in a CAD program helps to directly translate its shape into the computer model. Of course, you can recreate the design from scratch in CAD too, but using your sketch as the starting point will save time.
3. Work up your desired shape in an appropriate 3D modelling software or app. If you're not familiar with CAD there are a number of free programs that are easy to use and will get you started.
4. Ensure that the model is solid and closed. Also ensure that there are no duplicate faces or sides.
5. If your model has interlocking or moving parts, you'll need to allow a slight tolerance between these parts. Tim's Salvage Stool seat design has one tongue on the inside of each internal face that helps locate the three seat parts of the stool together. He offset these tongues 0.5 mm (1/50 in) from their corresponding grooves to enable them to sit together without jamming or binding.
6. If your print contains internal voids (most do, as a 3D print usually consists of a very thin outer skin, and a hollow, internal void that speeds up the printing process and saves on material), the void will need to be filled and strengthened with a lattice support structure. Your 3D printer may allow you to select how large you'd like the spacing of this internal lattice support structure. It's simple – the greater the weave, the weaker the structure. The closer the weave, the stronger the structure.
7. Select your desired material. Different materials provide different finish, strength and properties. Your choice of material will depend on what you intend to do with your print. For the Salvage Stool's seat, Tim chose sandstone, a gypsum-based resin, as this was the most cost-effective option. Strength was not critical, as it was being moulded again.

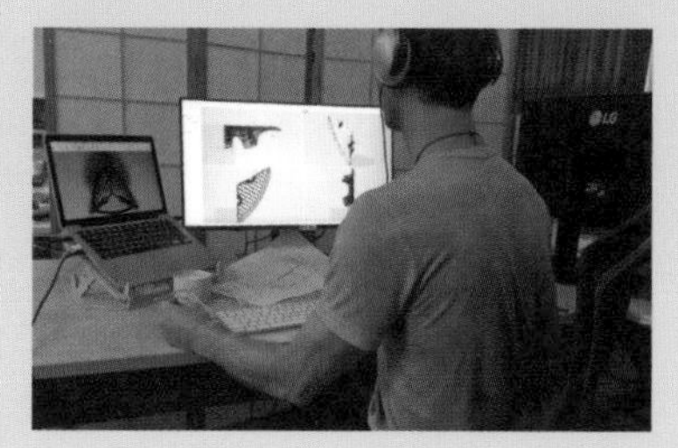

LPT Rooney

TRAVELLER STOVE

Welding

After years of experience in the humanitarian aid world, John Henderson set about developing a lightweight, robust and high-quality stove that could be sent out to disaster zones. The Frontier Stove developed a cult following, and John soon saw the need for a bigger stove; one designed for campers and hikers, tiny house, bell tent and container dwellers, and anyone living or working in a small space who wants to keep warm and cosy and be able to cook at the same time. The Traveller's lid doubles as a cooking hotplate, and can be opened up for open-flame cooking, while its side wings can be used as warming plates. While most wood-burning stoves are made of cast-iron parts, which are brittle and crack easily, this one is crafted from cut and welded steel parts, making the stove extremely durable, resistant to high temperatures, and long lasting.

MEASUREMENTS	Height: 46.5 cm (18⅖ in); width: 46 cm (18 in); depth: 42 cm (16½ in); weight: 24 kg (53 lb)
MATERIALS	Brass (handles), heatproof glass (pane), heatproof stove rope, mild steel (stove, door), stainless steel bolts
KEY TOOLS	Four-pound lump hammer, G-clamps, 2-lb rawhide mallet, 6-in tri-square, 12-in tri-square, spanners (wrenches)
KEY MACHINES	MIG welding plant, plasma cutting machine
TIME TO MAKE	6 hours (plus 2–3 days drying time)
LIFESPAN	Lifetime

JOHN HENDERSON AND NICK SHERRATT — Designer and welder/fabricator [Cornwall, UK]

John designs and prototypes every one of the camping products he develops at his camping technology company, Anevay, but it's welder Nick Sherratt, who completed a five-year apprenticeship with Rolls Royce back in the eighties, who makes the stoves that head out the door to customers, and ensures they are up to standard.

The pair work with an apprentice – Nick's son, Danny Sherratt – in a 4.6 × 9.1 m (15 × 30 ft) fabrication workshop in an industrial estate in Cornwall, England, with a converted shipping container, 6.1 m (20 ft) long, running off the side. Anevay also has a packing room and two-storey office onsite, and naturally, there's a Traveller Stove installed in the corner of each; perfect for keeping warm on chilly Cornish mornings.

Each stove takes the team about 5 hours to make. There's then another hour of sandblasting, painting and assembly. They take 2–3 days to dry, and then they're ready to ship.

Nick begins a new batch of stoves by getting all the parts together and checking them for quality. This includes steel tubes for the bodies, and component parts for the doors: heatproof glass panes, brass knobs for the handles and heatproof stove ropes around the openings, to make a good seal. He makes sure the profiles for the wings, door, base plate and other stove parts that have been laser cut offsite are up to scratch, then he dons his safety gear – overalls, welding mask and leather gloves – and gets to work.

He uses a plasma cutting machine and jig to plasma cut the door shapes out of the body tubes, assembles the rest of the parts and then welds them together. By now, the stove is beginning to take shape.

Nick says the door assembly is his favourite part of the making process. The door is bolted together and the stove bolted to its hearth tray with stainless steel bolts.

'The door has lots of intricate parts that need thinking about and putting together carefully. Everything has to line up perfectly or it just doesn't function. There's lots of skill to it.'

Danny stamps the identification number into the wings and the body with a lump hammer. Then Nick uses G-clamps to clamp piece parts together while they're welded. He uses a rawhide mallet to beat the glass brackets on the door into shape, using a tri-square to ensure that the lid sits straight on top of the body tube while it's welded together. Finally, Nick uses a spanner (wrench) to tighten the legs of the stove to its base plate, so it sits securely.

The stoves are then sent offsite to be sandblasted. When they arrive back at the workshop, the stoves are moved to a spray booth and set up ready for painting. Anevay's paint technician, Uldis, paints the stoves using a special heat-resistant stove paint. The stoves are left to dry for 2–3 days, depending on the colour of the paint; green, black or brown.

When the paint has dried, the stoves are assembled and carefully packed for shipping.

'We make our Traveller Stoves to last for years – a lifetime, if they're taken care of. We believe in repairing and replacing parts that wear out, rather than throwing things out and buying more. We try to make all the component parts available on our website in case people lose or break them, and we will often hand make parts on request,' says John, who designed the Traveller specifically for small spaces.

'Small spaces need a small footprint, so I made the cylinder vertical. The design I came up with stays lit overnight, can be cooked on, and has a glass pane, not just for aesthetics but also so that people can keep an eye on the fire.

'That's something I really took away from my work in disaster zones; everything needs to be multipurpose. Fires are essential for heat, but they must be able to be cooked on.'

“

We make our Traveller Stoves to last for years – a lifetime, if they're taken care of. We believe in repairing and replacing parts that wear out, rather than throwing things out and buying more.

CARING FOR YOUR TRAVELLER STOVE

Empty ash from the stove as it fills up: how often you need to do this will depend on what type of fuel you're using and how long you use the stove for. A good-quality, dry seasoned timber or heat log should require you to empty the ash out every few days. Always use a metal pail to empty ash, just in case any embers are glowing.

The stove should be rubbed down with an organic food-safe oil periodically, and especially before storage; anything you have to hand in your kitchen will do the job. Make sure the stove is dry before storing it away. Cooking splatters and splashes can be wiped down with a damp cloth when the stove has cooled.

Travellers aren't designed to be left out in the elements for long, but if they do get rained on, any wear and tear can be sanded down and touched up with heatproof stove or BBQ paint.

Temperature control tips:
Use the air control vents to adjust the temperature depending on what you're doing. On first lighting the stove, keep all the vents open as the fire takes. You can then shut the top vent halfway to keep the stove on tick-over, which is a great setting for keeping a room warm. If you're boiling water, open the vents up to allow lots of air in, which will get the fire roaring.

HOW TO

MAKE A PERFECT FIRE

Trust the innovative folk at Anevay to think outside the (fire) box. John and Nick advocate for a top-down fire instead of a more traditional bottom-up one; this means that, instead of starting with tinder and kindling on the bottom, then adding logs, you start with the biggest pieces of timber on the bottom and pile smaller and smaller pieces of kindling on top. They reckon starting a fire this way stops it from smoking and makes it more stable.

MATERIALS

- Dry, seasoned timber
- Kindling – small pieces of wood of varying size, from pencil-size to larger, 5 cm (2 in) thick pieces
- Firelighters or other dry tinder

TOOLS

- Matches or a lighter

METHOD

1. If you're lighting your fire in a wood burner, open all the air vents.
2. Line up a few small pieces of timber. On top of that, add a few smaller pieces of wood or kindling going the opposite way, so you're building a cross-hatch pattern with each layer. Make sure to leave a gap of around 1 cm ($\frac{2}{5}$ in) between each piece of timber, as this allows for a good flow of air and results in a better burn.
3. Add another layer of wood on top, using the smaller kindling pieces this time.
4. Place a couple of firelighters on top. John and Nick use natural firelighters made from wood shavings soaked in wax.
5. Now the fun part: light it, close the door and watch it come to life.

PRO TIP

- *If your fire is struggling to take, open the door to the stove slightly to allow a little more air in.*

Step 1

Step 2

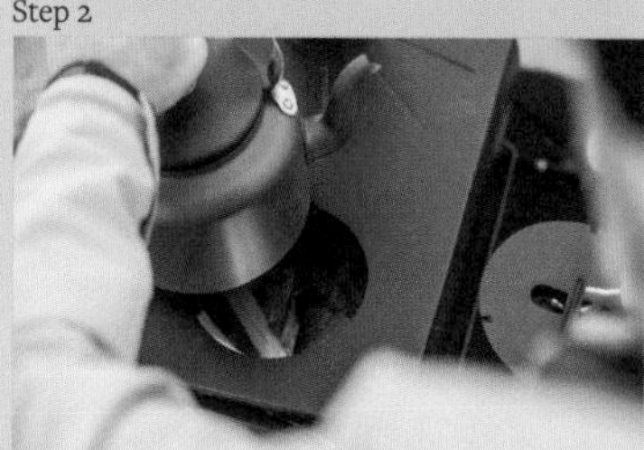

Step 3

PORTLAND RAZOR CO.
OR

THE HYDRA STRAIGHT RAZOR

Bladesmithing

In your grandfather's day, straight razors were the principal method of shaving. But over time the safety razor, with its dastardly disposable blades, and then the lazy-making electric razor, pushed the straight razor from its perch. But in recent years, championed by men who can (and want to) handle a blade, the straight razor is making a comeback. This razor was handcrafted by Scott Miyako and a crack team of makers in Portland, Oregon, who are committed to restoring the straight razor to its former glory. They also make and supply strops, run workshops on how to maintain straight razors, and can teach you how to wet shave without drawing blood. Mostly, they believe that the straight razor demands respect and responsibility, and that if you treat your blade well it will treat you well. Damn straight.

MEASUREMENTS	Length: 15.9 cm (6 in); razor width, including scales: 1.3 cm (½ in); razor height, including scales: 2.5 cm (1 in); weight: 85–113.4 g (3–4 oz)
MATERIALS	O1 tool steel, brass pins, brass washers, oil, Oregon Black Walnut, wood glue, polyurethane finish, Thermark
KEY TOOLS	Ballpein hammer, checkering files, round files, sharpening stones
KEY MACHINES	Belt grinders, buffing wheel, drill press, drum sander, kiln, laser cutter, random orbital sander, tempering oven
TIME TO MAKE	4 hours
LIFESPAN	Heirloom quality

SCOTT MIYAKO — Bladesmith [Portland, USA]

Scott Miyako has always been into making and fixing things. As a kid, he built a koi pond in his backyard, a skate ramp in the driveway and a recording studio in the garage. A qualified mechanical engineer, he doesn't feel right if he doesn't have a project on the go. Straight razors might just keep him on the straight and narrow for life.

'I've spent most of my life hand-working metal and wood. More than anything, I have a great appreciation for tools and finding the right tool for the job. A few years ago, my search for the perfect shaving tool led me to straight shaving. It quickly became a passion and I decided to bring straight shaving to others,' says Scott, who schooled himself in the art of making razors via online tutorials.

Today, Scott's company, Portland Razor Co., employs four people and turns out around 1000 razors a year. Scott and his team work in batches of four, crafting razors in a 112 sq m (1200 sq ft) workshop in an old building on Lair Hill, just south of downtown.

'The building was built in 1913 and the Portland Aerial tram runs right outside of our big windows. We are at the back of the building and look onto a tree line of Doug Firs on the ridge of the West Hills of Portland.'

Making a razor takes around four hours, not including the time spent waiting for heat treat cycles and glue/finish to dry. Apart from the initial blade profiling, everything is done in-house.

First comes initial grinding with a coarse-grit ceramic belt on a belt grinder. 'The steel is still soft at this point; we're removing material at a very fast rate. We generate lots of heat and sparks.'

Next, hand filing details into the straight razor. This includes decorative file work, notches near the heel and the point of the blade, and jimps (slender cuts or cross-hatched patterns to improve grip) along the tang (the unsharpened metal behind the hollow grind that serves as a hold point), and the spine. Then it's time for heat treatment.

'We normalise the blades in a kiln to reduce the stress and grain size in the steel. Then we quench them in oil and let them cool. At this point the steel is very hard and very brittle; to bring toughness back, we put the blades through two tempering cycles in the tempering oven.

Next, Scott grinds the surfaces of the blades to remove the decarburised metal from the outside of the razor until the blade is nearly sharp.

'Finer and finer grit belts are used until the desired polish is achieved. Extra care is taken to make sure that the razor stays cool during grinding, so that the temper (hardness) of the blade is not compromised.'

'We mark the blades with Portland Razor Co. branding by burning Thermark, an ultra-durable compound, onto the blades using our laser cutter. This is probably the most old school meets new school part of our process.'

The razors are given a high polish on soft muslin buffing wheels, and then it's time to prepare the scales and handles. Scales are the wooden part of the razor that protect the blades. Scott cuts them on a laser cutter, then glues, shapes and finishes them. Then he joins the three parts – the blades, scales and handles – together.

'We insert a brass pin into the pre-drilled holes of the blades and the scales. We place washers over the pin and flatten the end of the pin with a small ballpein hammer to hold the razors firmly inside the scales.'

Now it's time to put the cutting edges on the razors. Scott sets the bevel on a coarse diamond stone, then gets to work honing the razors on finer and finer stones until they reach their zenith on a 12,000 grit Japanese water stone. Lastly, the razors are stropped and coated with a light oil and packed into a protective box.

'When creating razors from raw materials, I feel happy. I believe the act of taking an otherwise useless object and turning it into something that is useful is one of the simple joys that we can have as humans.'

CARING FOR YOUR STRAIGHT RAZOR

Keep the blade free of moisture and oiled to prevent corrosion. Strop it (straighten and polish it) before each use, and have it honed by a professional bladesmith or sharpener regularly to maintain the health of the blade. Depending on the coarseness of your hair and how often you shave, honing should be performed every four to six months.

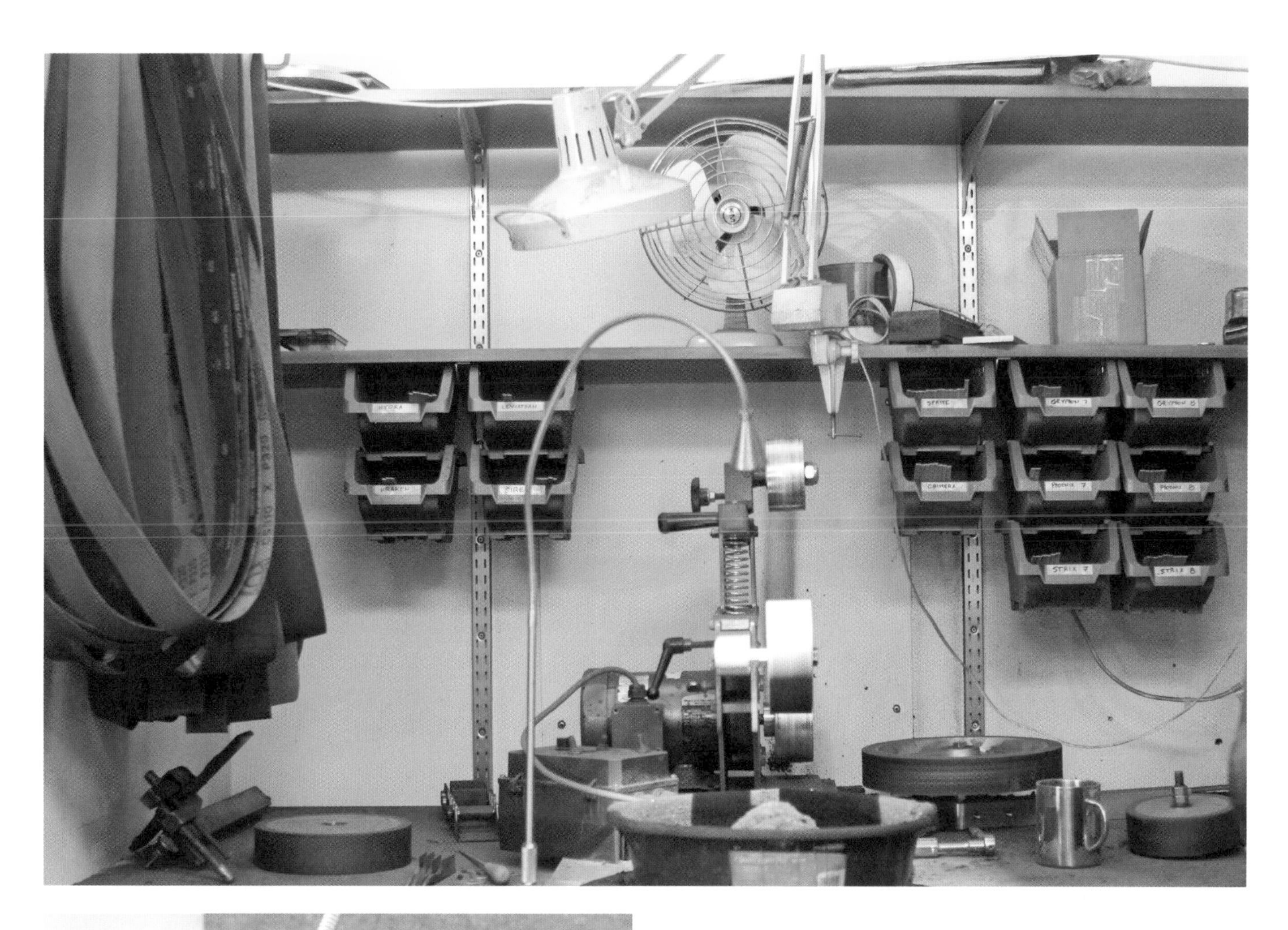

“

When creating razors from raw materials, I feel happy. I believe the act of taking an otherwise useless object and turning it into something that is useful is one of the simple joys that we can have as humans.

HOW TO

STROP A STRAIGHT RAZOR

MATERIALS

- Leather strop
- Straight razor

PRO TIP

- *When you're first learning to strop, be patient and don't try to work too quickly. A slow and proper stropping stroke is better for your blade than a quick but poor stroke. It's possible to damage your blade with improper strokes, so take your time to work up your technique. Once you have the stropping stroke down, this process will only take a matter of minutes.*

METHOD

1. Use the lacing at one end of the strop to tie it to a fixed anchor point, such as a towel rail, a hook on the wall or a closed and secure door handle.
2. Grip the strop's handle with your non-dominant hand, position your body just to the side of the strop, and pull the strop with just enough force to make it flat in front of you.
3. Now use your dominant hand to position the blade so that the spine and the edge lie flat on the stropping surface. The edge will be closest to you; the spine furthest. The spine should never leave the stropping surface. Don't roll the edge towards or away from yourself, or allow the strop to sag or become slack.
4. Push the blade away, keeping the spine and the edge flat on the stropping surface the entire time. The blade should glide across the stropping surface with little to no downward pressure. As you approach the end of the strop, stop the forward motion and flip the edge of the blade over the spine by rolling the tang in between your fingers. Complete the flip by gently laying the edge flat on the stropping surface. Now the spine will be closest to you, and the edge furthest away.
5. Pull the blade towards your body, stopping at the end of the strop.
6. Repeat this motion, beginning on the prep side. Strop 20 times on the prep side (one complete strop is a push out and a pull back). Complete your shave preparation on the finished side of the stropping surface. Strop 60 times.

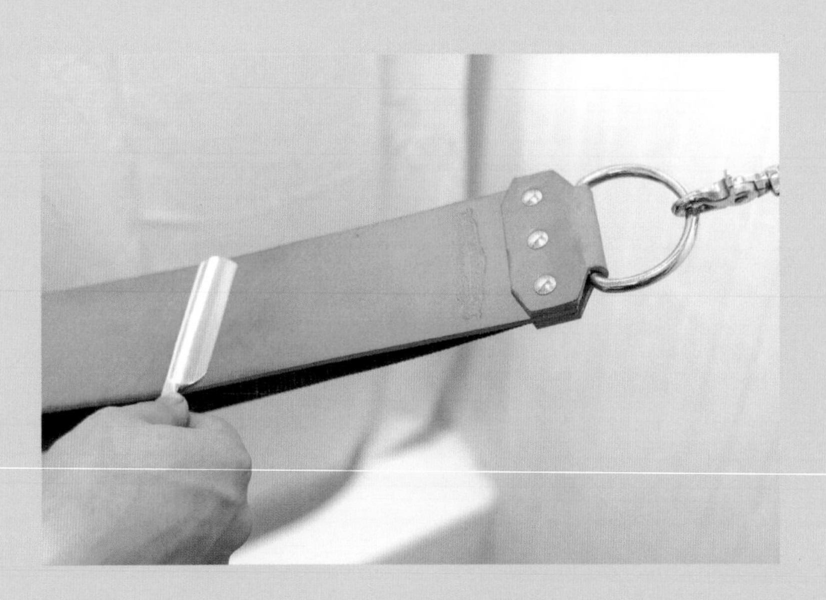

HOW TO

SHAVE WITH A STRAIGHT RAZOR

MATERIALS

- Shaving soap or cream
- Warm water
- Pre-shave oil
- Aftershave
- Alum block or styptic pen
- Post-shave balm

TOOLS

- Straight razor
- Shaving brush (badger, boar or synthetic)
- Strop
- Bowl for lather
- Towel

METHOD

1. Start by inspecting your razor and soaking your shaving brush in warm water.
2. Soak the hair you will be shaving until soft, either in a hot shower or with a hot towel. You may decide to apply a pre-shave oil at this point to lock in moisture.
3. Strop your razor.
4. Use your shaving brush to make a lather out of the shaving soap or cream by whipping them together in the bowl, as though scrambling an egg. A thick lather with fine bubbles provides the best barrier between the blade and your skin. Remove moisture by wringing the brush out or add moisture by dipping it in water to achieve the desired consistency.
5. Apply the lather to your face using circular strokes or paintbrush-like daubs. The lather should form soft, stiff peaks when you remove the brush.
6. Open the razor and hold it by the tang, with two fingers in front of the scales and two behind, and your thumb underneath.
7. With your other hand, pull the skin where you want to shave taut.
8. The First Pass, with the grain: shave in the direction of hair growth at a 15–30 degree angle off your skin. Use short, light strokes. Scott recommends starting with your cheeks, and progressing to more curved parts once you feel comfortable handling the razor.
9. The Second Pass, across the grain: re-lather and shave once again, this time moving across the direction of hair growth, or 'across the grain'. Think of this as hair reduction, rather than hair removal. Using lighter strokes will result in less irritation, and only shaving where you have lathered will make it more comfortable.
10. For an especially close shave, perform an optional third pass against the grain.
11. Once you are satisfied with your shave, rinse off any leftover lather, pat dry, and apply a few drops of aftershave to your skin. This helps clean, cool and tighten the skin. If you have any nicks or cuts, wet them with cold water and apply an alum block or styptic pen to disinfect and stop any bleeding. If you have sensitive or dry skin, apply a moisturising post-shave balm.

Step 4

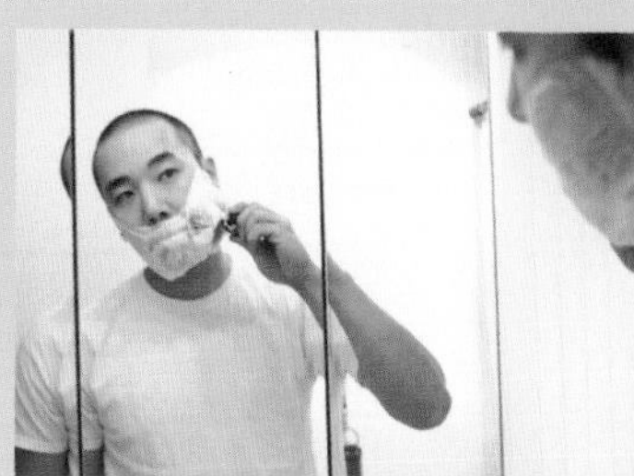
Step 5

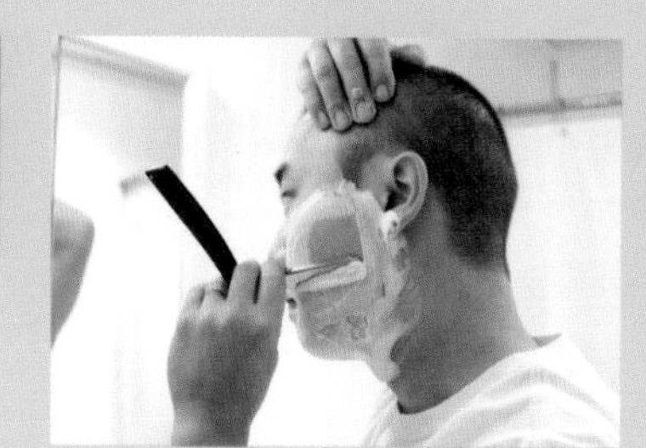
Steps 7–8

GREY THREE-PIECE WINDOW PANE CHECK SUIT

Tailoring

There's an old joke that a good tailor will get to know your body better than your wife or lover, and in many ways, it's true. A bespoke suit fitting is an intimate affair. Your tailor will want to hear about what you plan to do in the suit, measure you top to toe, and observe your body shape and posture. This snappy three-piece suit comprises a jacket (single-breasted, two buttons, peaked lapels, slanted pockets with flaps); waistcoat (single-breasted, six buttons, peaked fronts); trouser (classic cut, plain-fronted, buckle side adjusters on the waistband). It was handmade by tailors Robert and Daniel Jones of Sydney, whose bespoke suits are crafted using a full floating canvas construction. This means that canvas lines the interior of the jacket and waistcoat. It's a hallmark of quality that helps the suit drape well on your body, and ensures it will last.

MEASUREMENTS	Made to measure
MATERIALS	Cotton thread, lightweight 240g (8½ oz) Italian-woven super 150s pure Australian wool, unpolished horn buttons, wool and horsehair inner canvassing and linings, YKK metal zippers
KEY TOOLS	Measuring tape, sewing needles, small thread clippers, tailor's chalk, tailor's shears, tailor's square, thimble
KEY MACHINES	Buttonhole machine, pressing machines, sewing machines
TIME TO MAKE	50 hours
LIFESPAN	20 years +

DANIEL JONES — Bespoke tailor [Sydney, Australia]

Daniel Jones was 19 years old when he joined his father, Robert, in the family tailoring business. Daniel is the third-generation tailor in his family. His father taught him the art of cutting in the traditional English style, but over the years, Daniel has developed his own, more modern approach to making bespoke men's suits. This shows in his awareness of international fashion trends, and his social media presence.

The first step in making a suit is meeting with the client. This meeting lasts for as long as it needs to; Daniel likes to learn as much about his clients as he can. After the meeting, a pattern is made and cut.

'The pattern will be drafted onto the chosen cloth and our cutter cuts the suit to the requirements. The inner canvassing and linings are also cut at this stage to ensure all pieces fit like a puzzle. Then we select the right shoulder padding and under collar for our client's shoulder and neck shape,' says Daniel.

'It's important to get the initial cutting right as this will affect the end result. There's no coming back from a poorly cut suit.'

Each piece of the pattern is then basted (loosely sewn) together to form the shell of the suit coat. It's sewn together by hand and then shaped by hand according to the client's measurements and their posture details. The client is then invited to come in for fitting.

'We take the suit back to the cutting bench re-cut it in line with the adjustments made at the first fitting. It's then sewn together, and the pockets and buttonholes are finished.'

This requires hundreds of stitches throughout the inner canvassing and for the buttonholes, and is the most time-consuming part of the suit-making process. It's fine, detailed work that requires patience and a steady hand.

Then it's time for second fitting, and any final adjustments. The suit is pressed and then there is a final appointment, where Daniel and Robert ensure that they – and the client – are happy with the suit. This is the part of his trade that Daniel enjoys the most.

Zink & Sons is a family business in the true sense of the word. Daniel's father, Robert, was 15 years old when he left school and started to learn the tailoring trade alongside his father, Bill, who purchased the business from its namesake in the early 1950s. Robert started out sweeping floors, and ended up a master cutter.

Back then, in 1960, there were more than 40 people working in the tailoring room; today there are less than ten. The store is located in its original 1920s abode, now a heritage-listed art deco building on Oxford Street in one of Sydney's busiest neighbourhoods.

'Our workroom has a heritage feel; I really like that. We have a mix of old and new sewing machines, pressing machines and cutting benches, and as you might imagine, we all dress exceptionally well for work. Bespoke tailoring is a very detailed job and requires a lot of concentration and patience. There is no room for error, so we like to keep the workroom stress free.'

Daniel treasures his large Wiss tailor's shears, gifted to him by his grandmother after his grandfather, Bill Jones, passed away.

'It's a bit of a cliché but they really don't make tailor's scissors like they used to. These ones aren't available anymore. I really enjoy working with these; they remind me that I'm carrying on a family trade and they keep me honest and humble in my work.

'I keep these shears wrapped in a cloth every night and have them sharpened once every few months. I'll never know how many suits my grandfather cut with them, but these days they cut approximately five suits a week.'

> Bespoke tailoring is a very detailed job and requires a lot of concentration and patience. There is no room for error, so we like to keep the workroom stress free.

CARING FOR YOUR SUIT

In an ideal world, a man about town would have a different suit for every day of the week. Dry-cleaning chemicals are harsh on fibres, so dry-clean your suit as little as possible; once a month at most. Spot clean any spills or stains as soon as possible with a damp cloth.

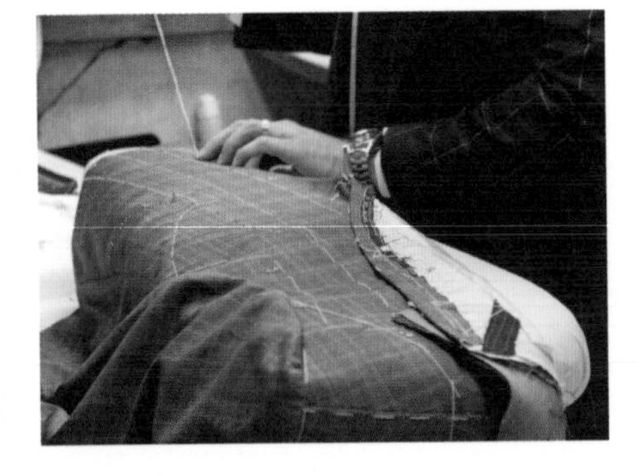

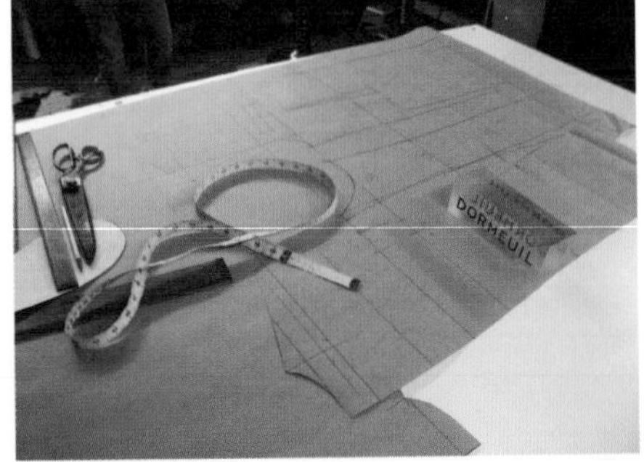

HOW TO

HEM TROUSER LEGS

MATERIALS

- Trousers
- Cotton thread

TOOLS

- Pins
- Tailor's chalk or dried soap block
- Tape measure or ruler
- Scissors
- Needle

MACHINES

- Overlocker
- Steam iron

METHOD

1. Put the trousers on, ensuring they are sitting at the correct height on your hips. Fold up the hem to the desired length. Daniel recommends having the back of the hem straight and just touching the heel, and sitting just on the shoe at the front. Insert pins on the fold both front and back.
2. Take the trousers off and lay them out flat. Mark the length where you placed the pins by drawing a line straight across the trouser leg with the chalk. From this line, mark another line 6.4 cm (2½ in) down each leg and cut the bottom line with scissors to get rid of the excess fabric.
3. Using the overlocker, stitch around the bottom of each leg. This will stop the fabric from fraying away. If you don't have an overlocker, using pinking shears at step 2 will do the trick.
4. Now fold the hem inwards to the first chalk line you marked and press the fold with an iron to create a sharp crease.
5. Turn the trousers inside out.
6. Thread the needle and stitch around the hem just below the overlock stitch, to secure the hem to the leg. When you get to the end, do a double stitch to make sure it won't come undone, then cut the thread and tie a knot.
7. Repeat on the other leg.
8. Now it's time to stitch the actual hem. Cut a piece of thread long enough to go around the hem twice. Thread the needle, then tie the two ends together so you are sewing with double threads.
9. With the hem folded up, start by making your first stitch on the inside seam. Do a double stitch here to secure the stitch. Then sew about 3 cm (1 in) in from the overlocked edge and pull it through.
10. Continue sewing around the hem using very shallow stitches, only picking up a few threads of the fabric so the stitches won't be visible from the outside. You will be only catching a few strands, but this will be plenty strong enough to hold. Make stitches every 6 mm (⅕ in) or so, sewing onto the overlocked hem then back onto the left part of the trouser fabric. Keep repeating this motion until you have made your way around the whole hem.
11. When you get to the end, do a double stitch, then cut the thread with the scissors and tie a knot.
12. Repeat on the other leg.
13. Turn the legs back to the correct side.
14. Give them one last press with the iron to remove any wrinkles or sewing tension that may have formed during sewing.

Step 2

Step 8

Step 1

BRDR. KRÜGER

TRIIIO TABLES

Furniture making

Think mid-century modern furniture and you think Denmark. And if you really know your tree types from your table legs, you think Bølling Tray Table – one of the Danish Modern movement's most recognisable furniture pieces. A collaboration between renowned architect Hans Bølling and seminal woodworking studio Brdr. Krüger, the Bølling Tray Table was released in 1963. Some 44 years later, the same design and maker collaborative wunderkind have released a new offering to the world of table lovers: the TRIIIO. Originally conceived by Hans Bølling in 1958, the design has been reworked for the 21st century, and enhanced with a smooth finish, floating tinted glass and brass accents. It comes as coffee, side or dining tables, all handcrafted in Brdr. Krüger's 3000 sq m (32,300 sq ft) workshop, 20 minutes by car from the Danish capital, Copenhagen.

MEASUREMENTS	Coffee table: Diameter: 90 cm (35½ in); height: 39 cm (15½ in); height glass tabletop: 10 mm (¼ in) Side table: Diameter: 80 cm (31½ in); height: 56 cm (23 in); height glass tabletop: 10 mm (¼ in) Dining table: Diameter: 135 cm (53.1 in); height: 70 cm (27½ in); height glass tabletop: 10 mm (¼ in)
MATERIALS	Beech, oak or walnut, brass plates, magnets, tinted and clear tempered glass
KEY TOOLS	Spokeshave
KEY MACHINES	Domino machine, stationary router, table saw, sanding machine
TIME TO MAKE	15 days
LIFESPAN	Heirloom quality

HANS BØLLING AND ALEXANDRE ARÉTHUSE — Architect and product developer/cabinet maker [Værløse, Denmark]

Brdr. Krüger was founded by brothers Theodor and Ferdinand Krüger in 1886. Now directed by Jyliane and Niels (fourth generation), and Julie and Jonas (fifth generation), the furniture-making studio works closely with its designers and architects from design conception through to finished product. The imprint of master craftspeople such as Alex Aréthuse can be seen in the perfect finish of each piece.

TRIIIO's design was realised nearly 60 years after Hans first drafted it, after Jonas asked to see Hans's archives. The design for the TRIIIO lay within, and was an exciting discovery – for Jonas and Hans, and for Alex, lead cabinet maker and product developer.

Alex's influence can be seen in the joinery, and in the innovative system of magnets that is used to affix and align the glass tabletop to the wooden legs and frame, enabling the table to be assembled, disassembled and stored. And his influence can especially be seen in the shape and finish of each and every TRIIIO table, which Alex crafts himself.

'It has been a great honour to work with Hans Bølling. He is an iconic Danish architect; he was a carpenter before becoming an architect, so he understands our work,' says Alex.

'He is also a very happy, kind and generous man who, despite being 85 years old, runs everywhere, talking to everyone and filling the room with positivity.'

To begin a table, Alex prepares the wood, taking it from its raw state to its working dimensions using a variety of machine saws.

CARING FOR YOUR TRIIIO TABLES

Keep the tables out of direct sunlight and away from heat sources like open fires or heaters. Keep the wood clean by wiping dust away with a dry cotton cloth. Clean the glass with a cotton cloth and soapy water or glass cleaner.

'The table is so simple and elegant, but in reality its organic shape demands complex, mathematically driven construction. There was a lot of measuring and calculating in the design phase, and I continue to be very careful with marking and measuring during dimensioning,' says Alex.

Construction involves ten distinct steps, and then comes shaping. Alex uses a hand router and various jigs and a spokeshave to transform the wood from its raw form to a beautifully finished organic shape.

He uses a domino machine and a stationary router to craft the joints – two for each leg. These enable the TRIIIO's three legs to be fitted together to form a strong and stable base for the glass tabletop. Once this is done, Alex assembles the legs.

Now Alex sands, dusts and prepares the wood until it is 'as smooth as a baby's ass' and then he applies the finish – oiled wax. With the beech, there is an ebony stain before the waxing; and for the smoked oak, a smoking process.

'I love to apply the first coat of the finish. At this point in the process the wood is very smooth, the touch and the smell are so good, and the finish reveals the beauty of the grain.'

Last comes the application of the glass tabletop, which has been crafted by a Danish company and is fixed to the table using the team's innovative magnet system.

'I glue three brass plates to the glass top with a transparent glue. I then fix a steel pin under those brass plates, matching it to its precise position in a hole at the top of each of the legs. I do this all by hand, with the help of precision jigs,' explains Alex.

'When the magnets and steel pins meet, the table top stays attached to the frame and can be lifted and moved without having to be detached – but can also be detached for ease of transport and storage.'

This clever feature also means that when the table is taken apart and reassembled it will be easy for the user to align the glass on the legs perfectly, according to Hans's design. Hans couldn't be happier with the outcome.

'It has been an immense joy to experience a sketch and prototype I made when I was 27 years old in the hands of the skilled craftsmen in Brdr. Krüger's workshop; to see them work their magic and witness how simple and elegant amendments have transformed my original sketch into three present-day tables with a timeless quality.'

"

I love to apply the first coat of the finish. At this point in the process the wood is very smooth, the touch and the smell are so good, and the finish reveals the beauty of the grain.

HOW TO

PREPARE WOOD FOR FINISHING

It's tempting to rush when preparing wood for finishing with stain, oil or wax at the end of a woodworking project. However, it's vital to the piece's aesthetic, feel and longevity that you take your time. Put on some music. Listen to a podcast. Get comfortable.

MATERIALS

- Wood

TOOLS

- 120 grit sandpaper
- 150 grit sandpaper
- Medium bristled brush
- 180 grit sandpaper
- Cotton cloth
- Lukewarm water
- Sanding foam 180
- 240 grit sandpaper

METHOD

1. If any machine or tool marks are left on the wood, begin with 120 grit sandpaper. If not, you can begin with 150.
2. Clean out the dust with a medium bristled brush.
3. Repeat sanding and cleaning with 150 then 180 grit sandpaper.
4. Dampen the cotton cloth in lukewarm water and wipe it over the surface to humidify the wood, then let it air dry for 10–15 minutes. You'll notice that the moisture will cause the wood grain to rise again; this is to be expected.
5. Now apply sanding foam and sand it with the 180 grit sandpaper; this will remove the raised grain.
6. Clean out the dust again.
7. Dampen the wood again, and let it dry for another 10–15 minutes.
8. Repeat the sanding foam application and sanding, then clean out the dust again.
9. Sand with 240 grit sandpaper and clean out the dust.
10. The wood should feel incredibly soft. It is now ready for finishing.

PRO TIP

- *When sanding, the movement of your hand should always follow the grain of the wood, otherwise you'll mark and scratch the wood. Between each step you should be able to feel the difference in the softness of the wood. Close your eyes and run your fingers over the wood to feel where more sanding is required.*

BEJEWELLED FOREST KITSUNE TATTOO

Tattooing

Tattooing is an ancient form of body adornment. Tattooed mummies have been recovered from more than 40 archaeological sites around the world. The oldest known tattooed dude is Ötzi the Iceman, who was discovered on the Austrian–Italian border in 1991 and dates back to between 3370 and 3100 BC. Ötzi had 61 tattoos – 19 groups of striking black lines spread across his body. This tattoo was designed by a tattoo artist and botanical illustrator who lives and works in the lush surrounds of the Dandenong Ranges, approximately 35 km (22 mi) east of Melbourne in Australia. It features a personalised version of a Japanese myth about a kitsune, *or fox. According to the myth, when a fox reaches 100 years old it can shapeshift into human form by donning a human skull.*

MEASUREMENTS	Approximately 25 × 35 cm (10 × 14 in)
MATERIALS	Antiseptic, stencil transfer fluid, tattoo inks (various brands: Dermaglo, Dynamic, Eternal, Fusion, Kuro-Sumi, Starbright)
KEY TOOLS	Drawing: coloured pencils, mechanical pencil, tracing paper, scissors. Tattoo: disposable tattoo grips, needle cartridges – 3, 5 and 7 needle outliners; 5 round shader; 5, 7 and 11 needle magnums
KEY MACHINES	Laptop, printer, photocopier, Inkjecta Flite tattoo machine, thermal stencil machine
TIME TO MAKE	20 hours
LIFESPAN	Lifetime

AMY C. DUNCAN — Tattoo artist [Dandenong Ranges, Australia]

When tattoo artist Amy C. Duncan was creating her tattoo studio, Artemisia, in 2015, she painted the walls a gentle shade of pink. Not a colour one might expect to see when walking into a tattoo studio, but one that Amy thinks makes the space soothing and feminine, and reduces any negative feelings her clients might experience.

‘Tattooing is painful and can induce feelings of stress and anxiety, so I wanted to create a space that would help reduce those feelings as much as possible. I filled it with plants, comfy couches, art and books that inspire my work, particularly natural history art. There’s a corner with an easel for painting, and a kitchen with every flavour of herbal tea,’ says Amy, who learned her art during a four-year apprenticeship under a tattoo artist named Shep.

Amy works at her studio five days a week. She begins her day with drawing or researching material for the day’s tattoos, and has her first appointment at 10 or 11 am. Once the client has approved the final design, Amy makes the stencil and sets up her work bench.

If she’s doing some colour on a piece she’s already started, Amy dispenses all the inks before she starts, and mixes up any custom colours required. Sometimes Amy’s day is split between two clients who get a couple of hours each, other times she has one client who sits all day.

This particular tattoo was done in 2016. It took 4 hours of drawing, 16 hours of tattooing across three full-day sittings, and contains 30 different-coloured inks, many of which Amy mixed by hand.

For the design, Amy consulted with her client, Kaye, about what she wanted and why before starting to draw with coloured pencils on tracing paper.

'The transparency allows for easy design modifications and the coloured pencil can be easily erased from the smooth surface without smudging.'

Kaye is a jeweller who loves Art Deco motifs, and wanted a bejewelled version of the Japanese *kitsune* concept. She also loves nature, and wanted a lush, green background including her favourite mushrooms and plants – toadstools, berries, oak leaves and acorns. She wanted the tattoo to be on her upper left arm, so Amy had to take the contours and movement of her arm into consideration when designing the tattoo.

When the time came to do the actual tattoo, Amy began by making a line drawing of the design on a second sheet of tracing paper using a mechanical pencil. She enlarged the line drawing on the photocopier to the correct dimensions and ran it through her thermal stencil machine to create the stencil on Spirit transfer paper. Then Amy cleaned Kaye's arm with Dettol antiseptic, applied a layer of stencil transfer fluid, and placed the stencil on her skin.

When she was satisfied that the stencil was in the correct position, Amy began the tattooing process. She used 3, 5, and 7 needle cartridges to complete the outline, then used 5 round, 7 magnum and 11 magnum needle cartridges to shade the darker areas. During the second and third sessions she added the brighter colours and highlights using the same needle configurations.

At the end of every session, Amy cleaned the tattoo with green soap, applied an antiseptic cream and covered it with plastic wrap to keep it clean and protected.

'Often the most rewarding part of the process is when I've completed a tattoo and both myself and the client are super stoked with the result. But sometimes, the end result isn't as important as the process of getting there. Maybe the best part is when I nail the concept for someone's tattoo when they were unsure of how to express their idea, or maybe it's when I experience a flash of inspiration during the drawing or research stage that informs future work. Inspiration can be very unpredictable and doesn't always show up when you want it to; it's a good day when it does.'

"

Inspiration can be very unpredictable and doesn't always show up when you want it to; it's a good day when it does.

CARING FOR YOUR TATTOO

Recommendations for tattoo aftercare differ from artist to artist, and from client to client. Amy recommends that you:

- Leave the initial plastic wrap dressing on for at least one hour, or up to four hours.
- Carefully remove the plastic wrap and wash the tattoo with a mild soap and warm water. A hot shower may sting a little at first, but is beneficial.
- Pat the area dry with a clean towel and allow your skin to fully air-dry as much as possible, for at least 10–15 minutes. If it's too cold to do that, a quick blast with a hairdryer set on low will dry your tattoo quickly.
- Apply a thin layer of your chosen tattoo aftercare cream. Only apply a small amount. (If you are ordering it online, make sure it will arrive in time.)
- Re-wrap your tattoo in plastic wrap and continue to repeat this cleaning and wrapping procedure for several days. Air your tattoo out between dressings if you feel you need to, but be careful not to get it dirty. If your clothes rub against your tattoo and it isn't wrapped, you'll need to re-apply the cream more often.
- Keep your new tattoo out of direct sunlight, salt water and chlorinated water for at least 10 days. Showering as usual is fine, but don't soak your new tattoo in the bath until it has healed.
- Your tattoo will start to peel within a week or so – don't pick or scratch at it! Your skin will be more or less back to normal within 2 weeks. You may like to apply moisturiser for a few weeks if your skin is still a little dry and scaly afterwards.
- Once it has healed, keep it out of the sun if you can and take good care of your skin by eating well and drinking water.

HOW TO

DESIGN AN IMAGE FOR TATTOOING

There's always a difference between the way a drawing looks on paper and the way it sits on the body, particularly for tattoos that wrap around arms and legs. Selecting an image and preparing it for tattooing takes time, care and practice. Here's how Amy does it.

MATERIALS

- Reference photos and books
- Photographs or tracings of the area to be tattooed
- Tracing paper
- Coloured pencils
- Mechanical pencil

TOOLS

- Camera
- Laptop or iPad pro + stylus

METHOD

1. Amy begins by assessing the image or idea to be tattooed for compatibility with the tattooing medium. Skin is textured, hairy, freckly and has pores and other irregularities, so is a vastly different drawing surface than paper. This means tattoos need a certain amount of contrast in them to look good. Features that tend not to work include too many tiny subtle details or shapes, long straight lines that will look distorted on the curves of the body, images with small white details surrounded by areas of black, or very small text that may become illegible over time. Images with shapes that can be delineated easily work best – Amy prefers not to do tattoos without at least some black outlines.
2. Once an idea has been established, Amy seeks out reference images. As Amy works mostly with naturalistic imagery, she tends to use photographs or natural history illustrations as references, and takes her own reference photos wherever possible. Sometimes an original drawing is necessary for a client's idea, other times an existing image is usable as is. Amy takes care to use images that are in the public domain if they will not be significantly modified for the final design.
3. Next Amy arranges the imagery so that it complements her client's body. Sometimes she does this digitally, by taking a photo of the body with her iPad pro and overlaying sketches and reference images, other times she sketches directly onto the skin with Copic markers and takes an impression of the sketch with clear contact paper. She then reduces the contact paper down to size on a photocopier and uses this image as a guide for sketches drawn on tracing paper. The imagery used needs to be scaled to the appropriate size for the body part, but also the appropriate size for the design – Amy wouldn't put a small, simple design on a thigh, for instance, or a large, complex design on a wrist. Amy takes into consideration what the client tends to wear and what parts of the tattoo will be visible in those clothes. Other considerations include musculature and body shape. Some parts of the body protrude and create highlights on the tattoo, while others are set back and will be somewhat less prominent. Amy aims to place the focal point of the design in a position where light falls upon so it can be easily seen, bringing cohesion to the design.
4. Once she's decided on composition it's time to draw the final design. The original sketch, whether digital or on tracing paper, is traced onto a second sheet of tracing paper and refined. Some parts are enlarged or reduced, some details and design elements are added or removed, and the whole design is tightened up.

HOW TO

DESIGN AN IMAGE FOR TATTOOING

5. Upon approval from her client, Amy traces over the image a third time using more tracing paper and a mechanical pencil to create a clean line drawing, or, if she is using the iPad, she creates a line drawing digitally on a separate layer and then prints it out. This line drawing is then photocopied to size and turned into a stencil to be applied to the skin.
6. Et voila, time for tattooing.

Step 2

Step 3

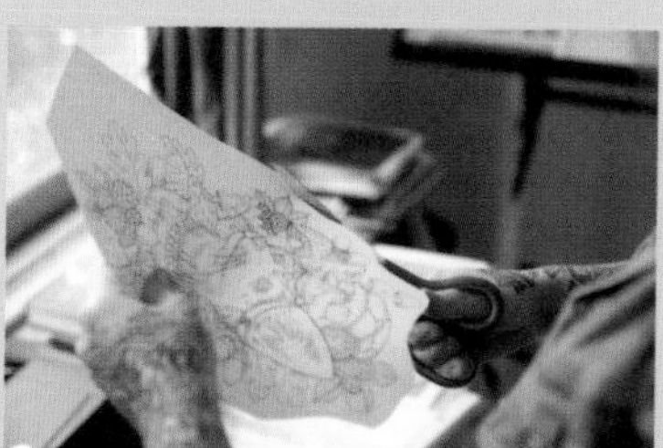

Step 4

Step 5

CUBE GLASS WHISKY TUMBLER

Glassmaking

There's this thing that's been irking whisky aficionados for decades. No, it's not the whisky vs whiskey debate, it's ... ice. Put ice in your whisky and it melts, slowly but surely filling your drink with water, which carries its own flavours, and messes with your drink. But you like your whisky chilled, right? So ice it is. Or ice it was, until glassmaker Nate Cotterman came up with the Cube Glass, a tumbler with a solid glass cube fused to its base. Made using traditional Venetian glass-blowing techniques that date back to the 13th century, the glass is designed to be stored in the freezer, ready to go when you and your whisky are. It'll keep your drink cool and its properties pure, and can be used for other tipples, too.

MEASUREMENTS	8.9 × 8.9 cm (3½ × 3½ in)
MATERIALS	Silica, potash, and quartz sourced from the Blue Ridge Mountains in North Carolina
KEY TOOLS	Blowpipe, diamond shears (cutters), jacks, marver, paddles, punty
KEY MACHINES	Furnace, glory hole, annealer
TIME TO MAKE	12 minutes
LIFESPAN	Forever (unless dropped)

NATE COTTERMAN — Glassmaker [California, USA]

Glass is intriguing stuff. It's hard enough to withstand high-speed winds, yet shatters with ease. It's made from opaque sand, but is incredibly transparent. It looks and behaves like a solid, but has the molecular randomness of a liquid. The first time glassmaker Nate Cotterman blew glass he found it hot, and awesome.

It was 2003, and he'd just begun his studies at the Cleveland Institute of Art, in his home state of Ohio. He was so intrigued by what he encountered at the glass shop that he never left.

'The heat feels good, the exhaustion after blowing glass gives me a great feeling of accomplishment. I'm creating an object from an amorphous glob of this beautiful, glowing material. It is truly unlike anything else I have ever experienced.'

These days, Nate operates out of a 372 sq m (4000 sq ft) workshop in El Segundo, on the Los Angeles County coast. Here, he designs and makes his own range of barware, home decor and lighting, and is earning an international reputation for his fusion of modern design with traditional techniques.

The workshop is kitted out with three furnaces. One, simply called 'the furnace', is where molten glass is kept, ready to go. The second is the glory hole, which is used to reheat a piece during the making process. The third is the annealer, used to slowly cool a finished piece without cracking or shattering it.

It takes Nate about 12 minutes to make a Cube Glass. The process involves four broad steps: casting the cube, blowing the cup, opening the cup, and inserting the cube. Each step demands thousands of minute actions and decisions.

'Each time I gather hot glass from the furnace I'm making adjustments to what I do to the material, how I move my body, when I blow, the level and rotation of the pipe, the amount of time I'm out of the heat versus in the heat ...'

Nate's tools are laid out on his tool rest with near surgical precision, from most used to least. 'Much of glass blowing is muscle memory, so having the tools in a specific location enables me to keep my focus on the object we are making.'

CARING FOR YOUR CUBE GLASSES

Rinse the glasses in room temperature water immediately after use. Never put them in the dishwasher, pour hot liquid into a cold glass, or freeze them with excess water left inside.

He starts by heating the tip of the blowpipe, then dipping it in the molten glass in the furnace and gathering it onto the end of the blowpipe like honey being swirled on a spoon. Nate cools and shapes the exterior of the molten glass blob by rolling it on a thick, flat sheet of steel called the marver. Then he blows air into the pipe to create a bubble inside the glass. If Nate needs to, he gathers more glass over the bubble to create a larger piece. Once the piece has been blown to its final size, the bottom is finalised using the jacks (large tweezers with two blades, used for shaping) and a cherry wood paddle. Then the molten glass is attached to the punty (a solid iron rod) and broken off the blowpipe. Nate heats and shapes the lip of the glass, then opens the cup. Finally, the lip is cut and polished.

Nate always works with an assistant because he needs to work so quickly, and remain so focused, that no time can be lost searching for or preparing a tool, or putting a piece down to open the glory hole doors. 'My assistant has to be very in sync with me; it's almost like a dance.'

Music, mostly rap, plays in the workshop for most of the day. Aside from safety glasses, Nate doesn't need any special gear. He simply dresses for the heat, in shorts, sneakers and a T-shirt.

'All stress and worry about outside life doesn't matter in the moment of making. I have to be fully engaged in an object, or it's not going to come out right.'

> "The heat feels good, the exhaustion after blowing glass gives me a great feeling of accomplishment.

HOW TO

MAKE A KILLER MARGARITA, NATE COTTERMAN STYLE

INGREDIENTS

- 60 ml (2 fl oz) tequila (Nate likes Herradura)
- 30 ml (1 fl oz) orange juice
- 30 ml (1 fl oz) freshly squeezed lime juice
- Splash of agave
- 1½ cups ice
- Fine sea salt, to serve

TOOLS

- Cocktail shaker
- Shallow bowl

METHOD

1. Chill two Cube Glasses in the freezer for at least 20 minutes before you make your drinks.
2. Combine the tequila, orange juice, lime juice, agave and ice in the cocktail shaker, and shake well.
3. Put the salt into a shallow bowl.
4. Remove the Cube Glasses from the freezer and wipe the rim of the glass with lime juice, then dip and spin the rim in the salt.
5. Strain the margarita into the Cube Glasses, and enjoy.

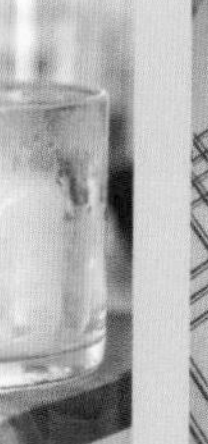

Step 1

Step 2

Step 4

Step 5

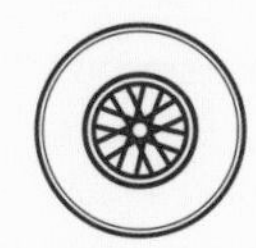

CANNON WHEEL

Wheelwrighting

The invention of the wheel is often cited as humankind's light bulb moment. The wheel dates back to 3500 BCE in Mesopotamia, when it was conceived for potters to throw pots on. It was another 300 years before some clever clogs thought of using them to get around on, and came up with the first wheeled vehicle: the chariot. Wheelwrighting is an ancient trade, and one that, in Britain at least, is going through a gentle resurgence. This wheel rolled out of Mike and Greg Rowland's workshop. The Rowlands' business is just one of two approved by Royal Warrant to provide wheelwrighting services to the British royal family. The father and son make and repair wheels and vehicles of all shapes and sizes using traditional timber, components and hot bonding techniques; this one is a cannon wheel made for a British military unit in 2012.

MEASUREMENTS	Diameter: 137.2 cm (54 in)
MATERIALS	Sapele (hub), Oak (spokes) and Iroko (felloes)
KEY TOOLS	Chisels, compass plane, draw-knives, hollow auger, planes, sledgehammer, spoke-dog, spoke-horse, spoke pointers, spokeshaves, traveller
KEY MACHINES	Band saw, lathe, morticers
TIME TO MAKE	3 days
LIFESPAN	25 years +

MIKE AND GREG ROWLAND — Wheelwrights [Devon, UK]

As its etymology suggests ('wright' comes from the Old English 'wryhta' – a worker or maker), wheelwrighting is a trade that dates back centuries. British wheelwrights Mike and Greg Rowland can trace their family's history in the trade back to 1331 CE. Greg learned his trade from his father, Mike, who learned it from his father before him.

The process of making a wheel begins in the middle, with the hub. Mike and Greg select perfectly dry, seasoned wood. It's roughly hewn, then measured, marked, cut and traced and trimmed to the correct circumference using a band saw. It's then finished using a pulley-powered lathe and chisel to turn the hub.

'Then we mark the hub with dividers and gauges, and form the angled mortices into which the oak spokes will fit by drilling and chiselling. These cannon wheels have twelve spokes.'

Next, tenons (joints) are cut for the spokes. Each spoke is placed in a spoke-horse, clamped, and then shaved to reduce mass, firstly by means of a draw knife and planes, then with a spokeshave (a small plane with a handle on each side of its blade, used for shaping curved surfaces).

'The spokes' feet are then driven with a sledgehammer into the hub. They each need to be precisely angled and aligned, but they also need to be tightly driven. Even though we're using sledgehammers, we have to be very careful not to split the spokes.

'The spokes are then finished with a spoke pointer and hollow auger to make a circular tongue, which will fit into the ash felloe – the shaped blocks that make up the rim of the wheel.'

In the past, wood for the felloes was often worked green (fresh and unseasoned) and adzed to get the necessary curve. As wood bending techniques developed, adzing diminished. Today, Mike and Greg cut the felloe with a band saw, then finish it with a compass plane to make a perfect circle.

'Each felloe is bored, ready for dowelling. The dowels hold the felloes together; we fit them into the spokes using a stick-like tool called a spoke-dog, which pulls the spokes together. Then we measure the wheel's circumference with a traveller (a circular measuring device) and add or subtract anything we need to – all the joints and curves have to be absolutely perfect. The best wheelwrights know when something looks right or wrong instinctively, without having to even measure.'

After that, it's time to cut and bend the iron for the tyre hoop that cradles the wooden wheel and sits against the ground. It's cut with a hacksaw and bent with a roller, and then the ends of the iron hoop are welded together. The wooden wheel is put on the bonding plate on the grass outside their workshop in the Devon countryside, ready for joining with the hoop.

But first the hoop is heated in a fire that Mike and Greg make nearby. When the hoop is cherry red, it's lifted out with tyre dogs and carried to the wheel – it takes both of them to do this. The hoop is quickly fitted over the wheel, and levered into place with the dogs, tampers and sledgehammers. This is called bonding.

'When it is in place, we cool the entire wheel by pouring cold water over it. The hot expanded metal contracts and shrinks with the cold, which pulls the wheel and all its component parts together.'

When the wheel is cold, Mike and Greg move into the final steps. They settle the tyre by hammering it to align it correctly. They drill and nail the tyre, then they chamfer it (cut away any right-angled edges or corners to make a symmetrical sloping edge) with chisels and spokeshaves. They sand the felloes and give the wheel a final clean, then finish it to the customer's final requirements with any painting or varnishing requested.

Bonding is Greg's favourite part of the process.

'There are a lot of hissing, snicking, spitting noises as the bonding happens. Within seconds, the steel tyre has transformed 20 or so pieces of wood into one solid working wheel using age-old physics. All the joints we've made just disappear. I get a rush from the history of what we're doing.'

Mike and Greg know they've done a good job when they never see a wheel they've made or repaired again.

'The wheels we make will last a hundred years or more – provided they are properly cared for. When we repair something, it's not just "mend something because it's broken", it's "why is it broken?" and "how can I prevent it from happening again in the future?"'

CARING FOR YOUR WOODEN WHEEL

Water is the enemy of a wooden wheel. If the wheel is going to be outside for any time, make sure it is turned frequently so moisture and weight don't sit in the same place. If the timber on the wheel is oiled, then a similar oil mixed 50/50 with a clear wood preservative should be applied at least once a year – the preservative helps to prevent rot and preserve the timber. If the wheel is painted, check it regularly for cracks, which will let water in. Any cracks should be touched up with fresh paint as soon as possible.

"

There are a lot of hissing, snicking, spitting noises as the bonding happens. Within seconds, the steel tyre has transformed 20 or so pieces of wood into one solid working wheel using age-old physics.

HOW TO

MAKE A MORTICE AND TENON JOINT

The mortice and tenon joint has been used for thousands of years by woodworkers around the world to join pieces of wood, mainly when the adjoining pieces connect at a 90-degree angle. It is simple and strong. In wheelwrighting, mortice and tenon joints are used to attach each of the spokes in a wheel to its hub (centre) and rim (exterior).

A basic mortice and tenon has two components: the mortice hole and the tenon tongue. The tenon is inserted into the mortice: a square or rectangular hole that has been cut into the corresponding timber. Here's how it's done, wheelwright style.

MATERIALS

- 2 pieces of timber to be joined
- 2 timber wedges

TOOLS

- Hammer
- Measure
- Mortice gauge (optional)
- Tenon saw
- Mortice chisel

METHOD

1. Mark the position of the mortice in the wood that will receive the tenon using a measure and mortice gauge. Exactly where the mortice should go will depend on what you're making, but a central location is advisable.
2. Place the second piece of wood in the vice, and mark the position and size of the tenon in the wood.
3. Use the tenon saw to cut the tenon, then remove it from the vice.
4. Now put the first piece of wood in the vice and use the mortice chisel to chop out the mortice.
5. Fit the tenon to the mortice.
6. Inserting the timber wedges either side of the tenon, wedge the tenon into the mortice and use the hammer to bang it firmly into place.
7. Use the tenon saw to trim the protruding tenon and wedges so that it sits flush with the mortice. Your join is complete.

Step 1

Step 5

Step 7

LITHIUM
GREASE
Ø = 3.6mm
= 0.141"
OS,M.

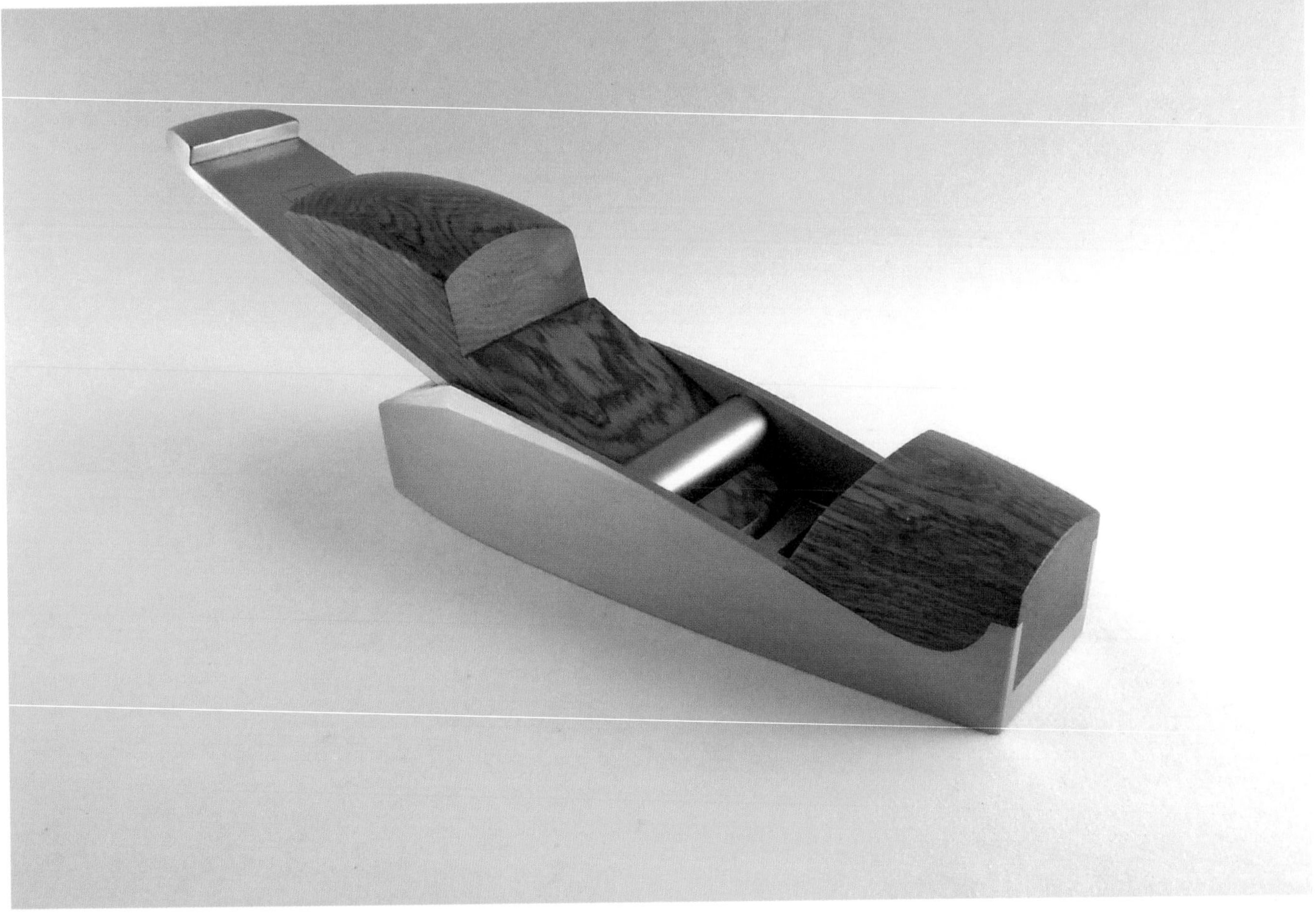

SLIPPER WOODWORKING PLANE

Toolmaking

If a toolbox were a bar, then infill hand planes would be the spirits that sit on the very top shelf. Named for the exotic hardwood that fits into the plane's metal shell, planes like the Slipper are considered collector's items – but that doesn't mean they don't get used. This one is ideal for fine trimming, flushing and spot smoothing tasks. It has a steel shell and rosewood infill with a carbon steel iron (blade), and a bed set to a 19-degree angle. It has been handmade in the workshop of Oliver Sparks, who couples contemporary design with traditional production methods, and does everything himself in a 69.7 sq m (750 sq ft) workshop that is heated by a 100-year-old wood burner, in a small makers' community on a farm in the English countryside.

MEASUREMENTS	Length: 12.7 cm (5 in); width: 4.4 cm (1¾ in); iron (blade) length: 3.2 cm (1¼ in); blade bed: 19 degrees
MATERIALS	O1 tool steel shell, Honduran rosewood infill, high carbon steel blade, wax oil finishing blend
KEY TOOLS	Metal scribe, brass template, chisels, clamps, diamond stones, files, floats, hacksaw, Japanese water stones, paper, pen, rasps
KEY MACHINES	Belt grinder, milling machine, muffle furnaces
TIME TO MAKE	4 days +
LIFESPAN	500 years +

OLIVER SPARKS — Plane maker [Leicestershire, England]

Oliver Sparks trained as a cabinetmaker, but it's hand planes, made the old-fashioned way, that have his heart. Apart from two digitally controlled muffle furnaces, there are no computers or automated machines in Ollie's workshop. He steers clear of digital technology in the design phase, too, preferring to develop his ideas using pen, paper and wooden mock-ups.

To begin making a plane, Oliver uses a metal scribe and brass template to trace the sidewall and sole shapes that will form the plane's metal shell and sole, and then cuts them out with a hacksaw.

The sole comprises two steel halves, which he cuts out on a milling machine and fits together with a tongue and groove joint. But the sidewalls are connected by means of dovetail joints.

'I make these by hogging out most of the waste using my vintage milling machine, then use a selection of fitters and engineer's files to refine the joints to their final shape. Then I use needle files and fitter's files to refine the metal dovetail joints and fit them into one another. To secure these joints and make them invisible, I hammer the protruding sections on my anvil. This deforms the dovetails into one another in an incredibly strong and stable union.'

Next, Ollie begins the wood infilling process. He makes templates from melamine sheet, using glue to temporarily stick them to the timber infill blanks.

'The templates provide a boundary for me to cut to using chisels, rasps and specialist plane-making tools called floats. Once the infills have been cut out and shaped I spend a few hours gradually fitting each one to its shell.

'When they are nearly airtight, I clamp them in place and drill holes for the rivets (metal pins to hold them together) through the body, sideways. I do some pre-polishing then apply industrial grade epoxy to the degreased body.'

He clamps them and leaves them to cure for 24 hours and moves on to the iron (blade). He marks it up, measures it and cuts it out then profiles it into shape with an industrial belt grinder.

Then Oliver fires up his two digitally controlled muffle furnaces. He heats the blade to 795°C (1463°F) in the first furnace.

'It comes out red hot, so I quench it in oil to harden the blade. It's too brittle to use like this, so it goes into the second furnace for a longer soak at 175°C (347°F). This tempering adds flexibility, but the heat-treating warps the blade slightly. It must be flat, so when it's cool I take it back to the belt grinder to polish the edges. Then to the surface grinder to grind both faces extremely flat, and true.'

The blade is sharpened on a mixture of diamond and Japanese water stones, then fitted to the plane body.

Next, Oliver positions the bridge – the metal bar that holds the wedge, which in turn holds the blade – in place, drills rivets into the plane's body, then fixes them by peining (striking them with a hammer). The body of the plane is now completed. Oliver cleans it up with a combination of belt grinders and hand filing, then adds the decorative chamfers (symmetrical right angles) and polishes them.

'I finish the plane by hand shaping the cured infills using rasps, files and chisels in very good light. I cut, shape and fit the wedge and then, when I'm happy with the shapes, I sand the wood smooth. I stamp my OSM touchmark onto the plane, and polish the infills.'

It's a long and involved process, but Oliver refuses to compromise on construction.

'I always choose the very best methods and techniques, regardless of how lengthy or difficult they are. If a detail takes twice as long but is the correct thing to do, then so be it.'

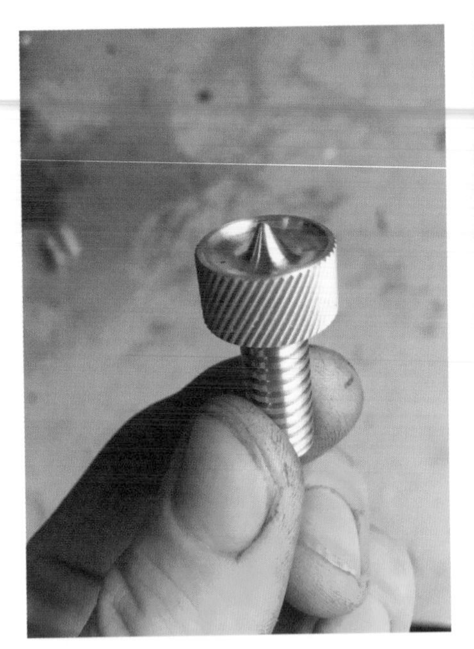

"
I always choose the very best methods and techniques, regardless of how lengthy or difficult they are. If a detail takes twice as long but is the correct thing to do, then so be it.

CARING FOR YOUR PLANE

For optimum performance, keep the blade sharp. Wipe down the steel parts to prevent rust after use, and wax the wooden parts once a week to keep the wood from drying out. When the blade gets too worn (this could take hundreds of years, depending on how much planing you do), have it replaced.

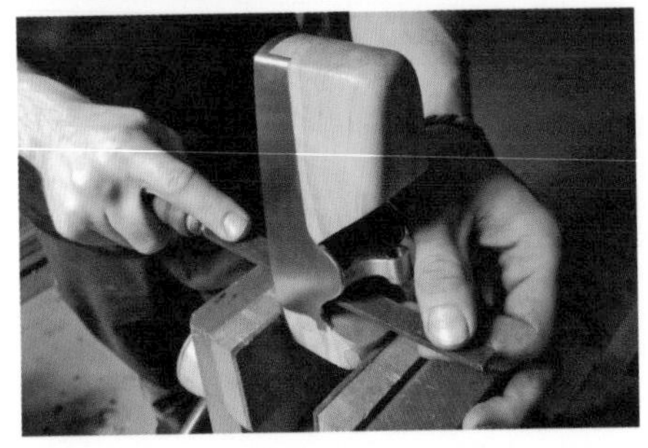

HOW TO

PLANE A PIECE OF WOOD TO FINISHED DIMENSIONS

A plane is your go-to tool for smoothing and shaping wood by hand. Used to shave thin, uniform strips from a piece of wood, planes remove what are called high spots, and create a smooth, level surface. Back in the day, all planing was done by hand; today, electric planing machines can do the job for you. But where's the joy in that? Here's how to do it yourself, the old-fashioned way – especially useful when you have one or two pieces of warped wood you want to repair, like cupboard doors or tabletops, but don't have the machines for the job. They'll likely be fine once you've planed off a few layers of the surface wood.

MATERIALS

- Wood to be planed

TOOLS

- Hand plane
- 220 grit wet/dry sandpaper
- Level or straight edge (optional)

PRO TIP

- *Avoid planing directly against the grain, as this can cause the blade to catch under minute, angled imperfections in the surface of the wood and cause tear out – literally, the tearing out of small, rough chunks from the wood's surface. This doesn't mean that you can't plane in multiple directions; sometimes you need to. Just don't plane against the grain.*

METHOD

1. Choose the right plane for the job. Hand planes come in different sizes and types; the defining characteristic of each is (broadly speaking) its size. The longer a plane's body, the more accurately it will straighten wood. However, shorter planes can be easier to control for precise detail work. 'Block', 'Jack', 'Jointer' and 'Smoothing' are some of the most common hand plane types.
2. Once you've selected your plane, sharpen its blade. It needs to be razor sharp before use – even brand new planes should be sharpened. To sharpen the blade, begin by placing a piece of 220 grit wet/dry sandpaper on a flat surface. Hold the blade at a 25–30-degree angle so that the bevel is flat against the sandpaper, then, maintaining this angle, rub the blade around the sandpaper in a circular motion, applying downward pressure all the while. When a burr (a tidy accumulation of metal shavings) forms along its back, the blade is ready to use. Remove the burr by wiping the back of the blade flat across the sandpaper.
3. Adjust the angle of the blade by turning the depth adjustment wheel (the small wheel just behind the blade assembly) until the tip of the blade protrudes just below the sole of the plane. This is important, as the angle of the blade dictates how thick the shavings you'll take from the surface of the wood will be. If the angle is too deep, you can jam the plane or tear your wood. Begin with a shallow angle, then increase the depth of the cut if necessary.
4. Now it's time to plane the surface of the wood. Place the plane at the wood's edge and begin smoothing and flattening the wood, applying downward pressure on the front knob and pressing forward with the back handle and pushing the plane across the surface in a smooth, continuous motion. Pay extra attention to any high spots or uneven spots on the surface of the wood.
5. If you are unsure how level the surface is, try using a level or straight edge to help you find uneven areas.
6. To fix any tear out, try re-planing the jagged spot along the grain of the wood or sanding it smooth.
7. Check the accuracy of your planing. Ideally, after you plane your wood, you'll have a smooth, flat surface that sits flush with any adjacent pieces of wood. Check your wood's flatness and smoothness by laying a straight edge along its surface. The straight edge should sit flush against the face of the wood. If there any gaps underneath your straight edge, this means your straight edge is sitting on a high spot, and it's time for more planing.

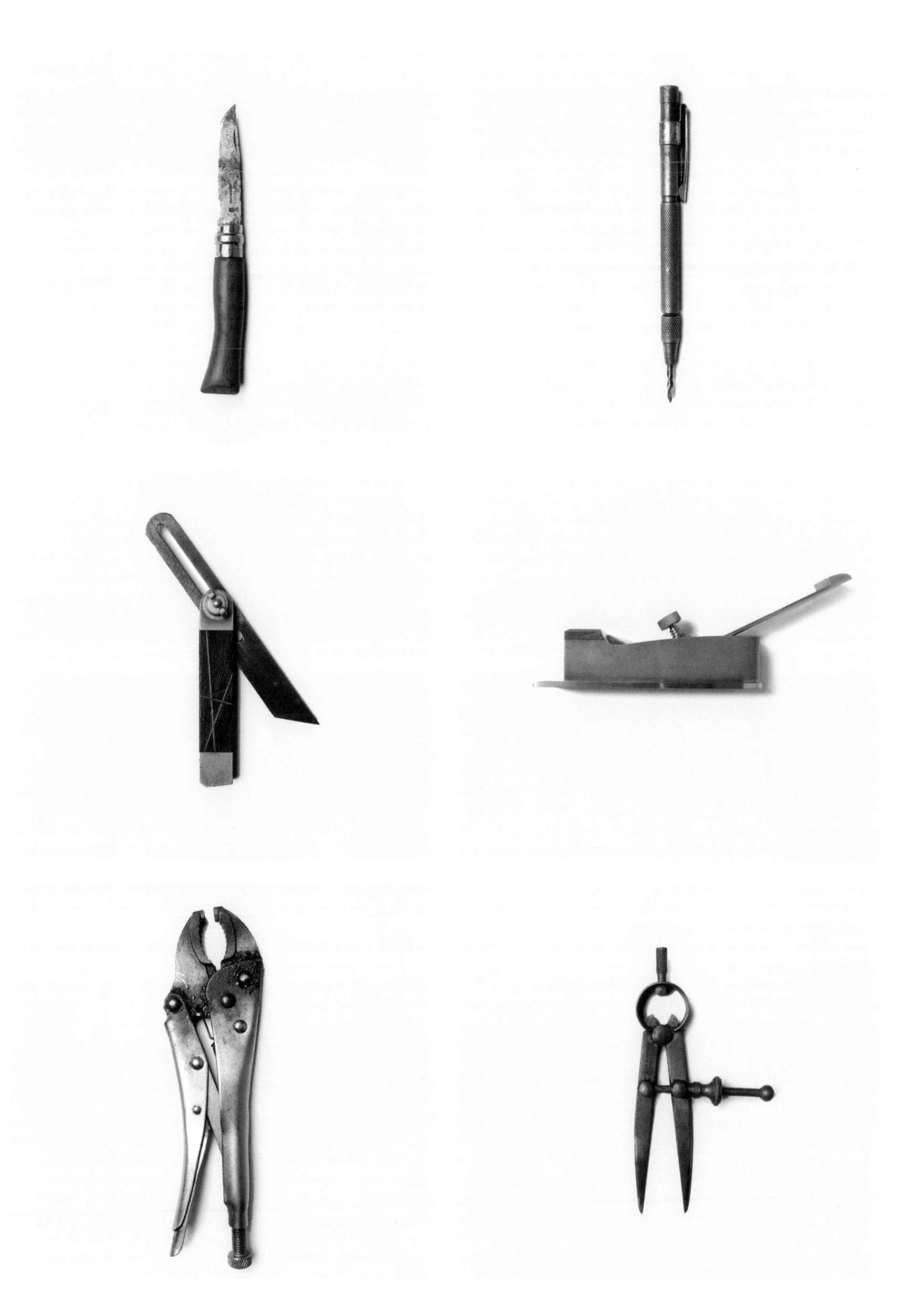

McGONIGLE
ENAMEL
Minute Repeater
McGONIGLE
CEOL
Ireland
Ref. McGJC15
Au750

McGONIGLE CEOL MINUTE REPEATER WRISTWATCH

Horology

There are wristwatches that tick miserably along until they just ... stop, and there are those that stand the test of time. This minute repeater, the Ceol, is one that will still be going when your great-grandchild puts it on. A minute repeater is a watch that has a mechanical complication that chimes the time on demand when the wearer activates a slide on the side of the case. The first examples appeared in the mid-18th century. This one was handcrafted in Ireland by a Swiss-trained horologist, and released in March 2016. With more than 283 moving parts, it's McGonigle Watches' most complex timepiece, and one of a limited edition of six. It has a power reserve of 52 hours, is water resistant to 30 m (98½ ft), and its ratchet (winding) system features dents de loup *(wolf's teeth) gearing, which is intensive on the hand-finishing front, but incredibly precise.*

MEASUREMENTS	Diameter: 41 mm (1½ in)
MATERIALS	Alligator skin (strap), sapphire crystals (front and back), enamel (dial), German silver (mainplate and bridges), gold (arrowheads on hands and buckle), 'jewels' (synthetic), mirror polished screws, steel (hands), white gold (case and crown)
KEY TOOLS	Burnishers, cutting and polishing broaches, eyeglasses, files, pushers
KEY MACHINES	CNC machine, electronic timing machine, lathe, milling machine, pusher, watchmaker's lathe
TIME TO MAKE	2 months +
LIFESPAN	Heirloom quality

STEPHEN MCGONIGLE — Horologist [Neuchatel, Switzerland]

Not many craftspeople assemble a product only to take it apart, clean it and then put it all back together by hand – but that's exactly what watchmaker Stephen McGonigle does with each and every one of his handcrafted timepieces. This painstaking assembly, disassembly and finishing process means a single watch can take more than two months to complete – and that's after months of design and prototyping.

'The first assembly is a process where each and every component is fitted and adjusted to the watch to ensure it is going to work. It takes weeks, and the mechanism is being exposed to dirt and dust the whole time,' explains Stephen.

The components also need to be finished, which means the edges of every single bridge, lever, spring and screw are bevelled and polished. This is done completely by hand.

'There would be no sense in doing the finishing during the first assembly, as there would be a risk of undoing the delicate work that is the finishing. During the second assembly every part is thoroughly cleaned and the necessary surfaces well oiled.'

The final assembly of a watch like the Ceol is a process tinged with relief. It's the culmination of months and months of work that begins with sketches, then 3D drawings in CAD, then making the first parts.

'We build a prototype, where we test the parts that are being made. Prototype work can be frustrating but also very interesting and exciting. Components often need to be adjusted by hand – for example, filed down, burnished, cut and polished – in order to fit with others, and to function. Sometimes the adjustments are so intensive that it's better to simply redesign and remake the piece,' says Stephen.

'Once we're happy with the prototype, we assemble the first piece, adjusting one component at a time as the watch is built up. Once the functions are correct and tested, the watch is completely dismantled. All 283 pieces are finished by hand, cleaned and oiled, and then reassembled.'

It's incredibly close and precise work. The hands of the Ceol alone, which are made from polished blue steel and tipped with white diamonds, take three days to make. Stephen spends hours and hours at his workbench, which he considers his most important tool.

'I spent a lot of money on it ... it's a beautiful thing and it's comfortable. It has beautifully sculpted armrests, which make long hours at the bench much easier.'

How does he maintain his focus? Sometimes, it's a struggle, he admits. An eclectic collection of music, from Rage Against the Machine to Macklemore and the Black Keys, to classical music and opera, is on hand to match his mood.

'Near the completion of a watch there is often a slight nervousness. Imagine that the watch is now perfectly clean. Every single component and every surface on every component has been finished. A lot of the steel work has a mirror polish. The tiniest mistake, the slightest slip and work that took days could be ruined. When the watch is finally complete and fully tested, there's a certain amount of relief.'

The view of deer grazing in the garden outside the window and of Lake Neuchatel in Switzerland and the Alps beyond pulls Stephen's gaze and calms his nerves.

Stephen studied his craft in Dublin, Ireland, from 1993–1996. After a year restoring antique and vintage watches, he moved to Switzerland.

The horologist's fascination with timepieces dates back to his childhood in country Ireland, where Stephen's father, Johnny, developed a reputation as a clock repair enthusiast. People brought clocks and watches to him for repair, and Stephen caught the horology bug. His older brother John, also a horologist and his business partner at McGonigle Watches, was equally fascinated.

'When I was restoring antique and vintage watches in London, I was surprised at the good condition of high-quality timepieces, despite their years. This was often down to the materials used and the production and finishing.

'When we started McGonigle Watches, we were determined to make watches that would last. This means using the highest quality materials, and preparing and finishing them to the highest standard.'

CARING FOR YOUR MECHANICAL WATCH

A mechanical watch is tough, but it's not a rough and ready watch that can be worn around the clock. Make sure you take it off to go skydiving, or whitewater rafting, for example. Have your watch serviced by a qualified horologist every four or five years. A service entails cleaning and oiling the mechanism, and a full control (the same checks and tests that are carried out when the watch is first completed, for functionality, waterproofing, timing and so on).

Wind the watch regularly. It's not possible to overwind a mechanical watch. In fact, it's better to keep it running than to let it run down. Simply wind until you feel a strong and sudden resistance; that's the mainspring, telling you it's fully coiled.

> Near the completion of a watch there is often a slight nervousness ... The tiniest mistake, the slightest slip and work that took days could be ruined.

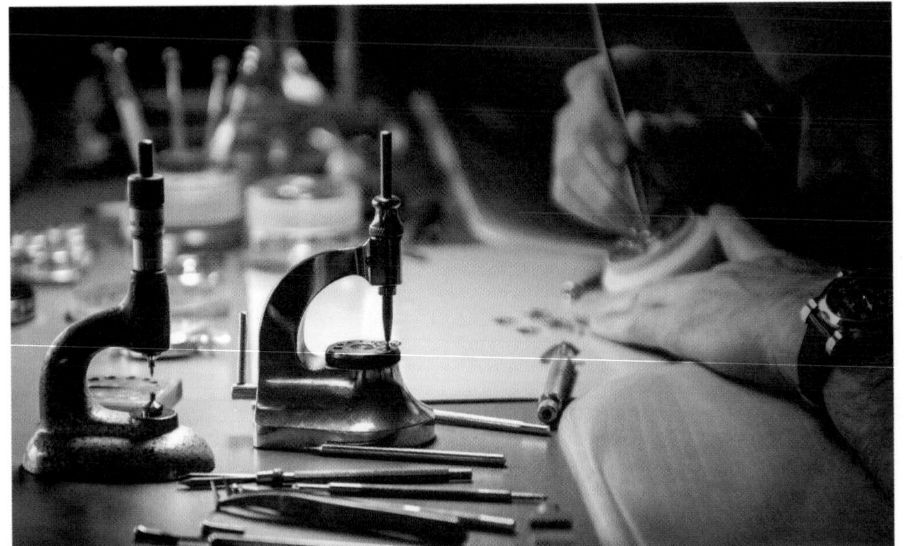

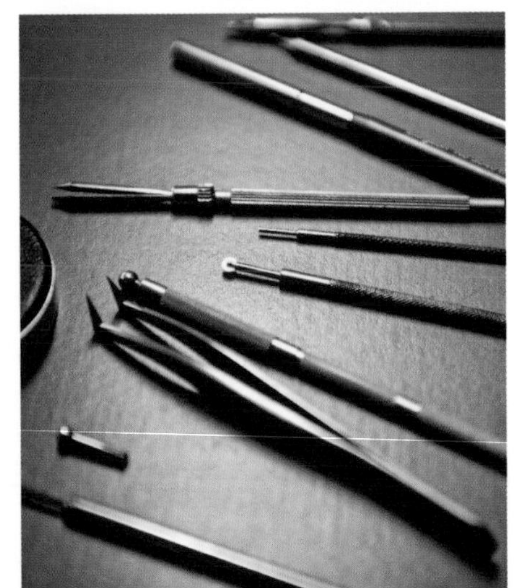

HOW TO

MAKE THE TRAIN BRIDGE OF THE CEOL

In mechanical watches, bridges hold all the parts in place. The train bridge is one of five in the Ceol. Like the others, it's screwed to the mainplate; the largest piece in the Ceol's mechanism and the foundation for the watch. It can take Stephen up to five days to make the train bridge.

MATERIALS

- German silver
- Sapphire jewels
- Two pins
- Synthetic jewels (27, of different sizes)
- Pith wood from the Elder tree

TOOLS

- CAD software
- Lathe
- Milling machine
- CNC machine
- Diamond pastes
- Diamond paper
- Small hand motor with diamond burr

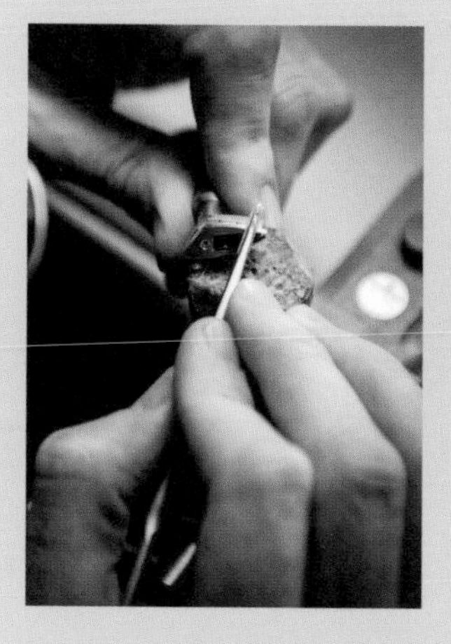

METHOD

1. After the design for the train bridge was finished, 3D construction in CAD software began. This is one of the most important stages, as a miscalculation here can prove very expensive and time-consuming. One of the incredible advantages of CAD technology is that each component's interaction with other components in the watch can be measured and assessed. It gives a good idea of what the real thing will look like.
2. German silver is the main material used in the train bridge. The part was machined. Stephen started off making single pieces by hand, using his lathe and milling machine. Once he was certain the component was good, he used a CNC machine, which can make parts quickly and accurately.
3. Next came prototyping. For the train bridge, Stephen friction-fitted (pushed together using a machine called a pusher) two pins into the train bridge, which position the bridge on the mainplate. The jewels, used as bearings for the train wheels, are also friction-pressed into place. Stephen adjusted the jewels to allow the wheels a small amount of play, typically just a couple of hundredths of a millimetre. The central part of the top surface of the jewels is concave. This is to accommodate oil, which reduces the friction between the jewel and its bearing surface; in this case, the pivots of each wheel.
4. Once the prototyping was finished, the watch was disassembled and the finishing of each part began. The first thing to be done was the engraving. The design for the engraving was planned with all the bridges considered as one. It's very unusual to see the different bridges so close together but this was a deliberate design feature, which helps bring out the amazing engraved work. This engraving was designed by the artist Frances McGonigle (Stephen's sister) and done by Madame Rothen in Le Locle, Switzerland. It took more than a week.
5. Once the engraving was complete, Stephen went over the top surfaces with diamond pastes and diamond paper to remove scratches.
6. Then the bevelled edges of the bridge needed to be mirror-polished. The piece is machined with a bevelled edge, but there's a lot of work needed to achieve the high polish. As with all McGonigle's finishing, this was done by hand. The edge of the bridge also received a grainy finish, which contrasts nicely against the polished bevels.
7. Lastly, the countersinks, where the jewels are located, also needed to be polished. Much like the bevels, the countersinks are also machined at the CNC stage. This polish was also achieved using a small hand motor. A soft wood was filed to match the shape of each countersink and, using diamond paste, this wood was held in the hand motor and rotated at speed while pressed against the bridge.
8. Once cleaned thoroughly, the part was ready for reassembly.

THE ARTISANS

Shop apron | Randi Jo Smith, Randi Jo Fabrications | randijofab.com

Wolfpack backpack | Wade Ross Skinner, Wolf & Maiden | wolfandmaiden.com

Bamboo bicycle | David Wang, Bamboo Bicycles Beijing | bamboobicyclesbj.com

Chef's knife | Iain Hamilton, Mother Mountain Forge | mothermountainforge.com

Spalted beech bowl | Franz Keilhofer, Gingerwood | gingerwood.de

Stoneground wholemeal sourdough bread | Ken Hercott, Bread Builders Maldon | facebook.com/breadbuildersmaldon

Shoe Brush Lovisa | Åke Falk, Iris Hantverk | irishantverk.se

Eternal calendar | Shigeki Yamamoto| shigekiyamamoto.com

Blackwood rod back settee | Glen Rundell, Rundell & Rundell | rundellandrundell.com.au

Campaign chest | Jeremy Zietz, Jeremy Zietz Studio | jeremyzietzstudio.com

Novena Heirloom laptop | Kurt Mottweiler, Mottweiler Studio and bunnie Huang, bunnie Studios | mottweilerstudio.com and bunniestudios.com

Gubbeen Smokehouse Chorizo | Fingal Ferguson, Gubbeen Smokehouse | gubbeen.com

Bathtub Cut Gin | Adam Carpenter and Wes Heddles, Prohibition Liquor Co. | prohibitionliquor.co

Split cane fly fishing rod | Edward Barder, The Edward Barder Rod Company | barder-rod.co.uk

Warre Wraps | Emily Gimellaro and Tania Dickson | etsy.com/shop/warrebeeswaxwraps

Handcrafted conservation frame | Shaun C. Duncan | shauncduncan.com

Cast-iron frying skillet | John Truex and Liz Seru, Borough Furnace | boroughfurnace.com

Tube garden trowel | Travis Blandford, Grafa | grafa.com.au

The Galileo globe | Peter Bellerby, Bellerby & Co Globemakers | bellerbyandco.com

Primrose electric guitar | Mitch MacDonald, Primrose Guitars | primguitars.com

Waxed Cotton Camper cap | Otis James | otisjames.com

Fire cider | Brydie Piaf | cityhippyfarmgirl.com

Dive Helmet pendant | Orion Joel, Orion Joel Custom Jewellery | orionjoel.com

Ansfrid jumper | Sandrine Froument-Taliercio, Atelier Lune de Nacre | atelierlunedenacre.com

The Derby lamp | Re'em Eyal, Studio Oryx | Israel | studioryx.com

Persevere letterpress print | Jennifer Farrell, Starshaped Press | starshaped.com

XX motorcycle | Jason Leppa and Sean Taylor, Gasoline Motor Co. | gasoline.com.au

Union neon sign | Scott Adamson, Gaslight Electric | gaslightelectric.com

The Commuter pannier bag | Cathy Parry, Industrial Sewing Workshop | industrialsewingworkshop.com

Washi paper | Fuchikami-san, Kumano Washi | kumanowashi.com

McFadyen picnic rug | Les, Jian Ling Liu, Anthony and Kirstie Mananov, Otto & Spike | ottoandspike.com.au

Handmade concrete planter | Kristy Tull, Fox & Ramona | foxandramona.com.au

Ebb Tide pebble plates | Kim Wallace, KW Ceramics | kwceramics.com.au

7" and 12" vinyl records | Connor Dalton and Daniel Hallpike, Zenith Records | zenithrecords.org

Dapple-grey Stallion rocking horse | Olivia O'Connor | oliviaoconnor.com.au

Shell Cordovans | Antonio Garcia Enrile, Enrile | enrile.es

Cirus Street skateboard | Dániel Bolvári, Cirus Skateboards | ciruskateboards.com

Beard soap | Evan Worthington, Craftsman Soap Company | craftsmansoap.com

Wood Grain BoomCase sound system| Dominic Odbert, The BoomCase | theboomcase.com

Ring-necked parakeet specimen | Jazmine Miles-Long | jazminemileslong.com

Salvage Stool | Tim Wigmore, Designtree | designtree.co.nz

Traveller Stove | John Henderson and Nick Sherratt, Anevay | anevay.co.uk

The Hydra straight razor | Scott Miyako, Portland Razor Co. | portlandrazorco.com

Grey three-piece window pane check suit |Daniel Jones, G.A. Zink and Sons | zinkandsons.com.au

TRIIIO tables | Hans Bølling and Alexandre Aréthuse, Brdr. Krüger| brdr-kruger.dk

Bejewelled Forest Kitsune tattoo | Amy C. Duncan, Artemisia Custom Tattooing| amycduncan.com

Cube Glass whisky tumbler | Nate Cotterman | natecotterman.com

Cannon wheel | Mike and Greg Rowland | wheelwrightsshop.com

Slipper woodworking plane | Oliver Sparks, OSM | oliversparks.co.uk

McGonigle Ceol minute repeater wristwatch | Stephen McGonigle, McGonigle Watches | mcgonigle.ch

PHOTOGRAPHY

Shop apron | Eric Reichmuth

Wolfpack backpack | Stan Kaplan, stan-kaplan.com

Bamboo bicycle | Bamboo Bicycles Beijing, bamboobicyclesbj.com

Chef's knife | Mother Mountain Forge, mothermountainforge.com

Spalted beech bowl | Nadine Schachinger, Herzflimmern, herz-flimmern.com

Stoneground wholemeal sourdough bread | Rebecca Newman Photography, rebeccanewman.com.au | Bread by Sprout Bakery

Shoe Brush Lovisa | Iris Hantverk, irishantverk.se | Anna Kern, annakern.com

Eternal calendar | Shigeki Yamamoto, shigekiyamamoto.com | Alexa Hoyer for Tictail, alexahoyer.com | Rahel Centurier

Blackwood rod back settee | Glen Rundell, rundellandrundell.com.au | Fred Kroh, fredkroh.net

Campaign chest | Jeremy Zietz, jeremyzietzstudio.com

Novena Heirloom laptop | Kurt Mottweiler, mottweilerstudio.com | Scott Torborg, scotttorborg.com

Gubbeen Smokehouse Chorizo | Fingal Ferguson, gubbeen.com

Bathtub Cut Gin | Prohibition Liquor Co, prohibitionliquor.co | Vanessa Murray & Steven Manos, vanessa-murray.com

Split cane fly fishing rod | Richard Faulks, rfaulks.com

pg. 73 Fish image, **David Wei / Alamy Stock Photo**

Warre Wraps | Emily Gimellaro, Tania Dickson, Bianca Esther

Handcrafted conservation frame | Shaun C. Duncan, shauncduncan.com

Cast-iron frying skillet | Borough Furnace, boroughfurnace.com

Tube garden trowel | Grafa, grafa.com.au | Marnie Hawson, marniehawson.com.au

The Galileo globe | Bellerby & Co, bellerbyandco.com | Alun Callender | Tom Bunning | Ana Santl

Primrose electric guitar | Primrose Guitars, primguitars.com

Waxed Cotton Camper cap | Sionnie LaFollette, sionnie.com

Fire cider | Brydie Piaf, cityhippyfarmgirl.com

Dive Helmet pendant | Breeana Dunbar Photography, breeanadunbar.com

Ansfrid jumper | Atelier Lune de Nacre, atelierlunedenacre.com

The Derby lamp | Studio Oryx, studioryx.com | Lynn Counio, lynncounio.com

Persevere letterpress print | Starshaped Press, starshaped.com | Matt Rieck

XX motorcycle | Gasoline Motor Co, gasoline.com.au | Jayne Moberley, jaynemoberley.com

Union neon sign | Gaslight Electric, gaslightelectric.com

The Commuter pannier bag | Erica Lauthier

Washi paper | Kumano Washi, kumanowashi.com | Bridget Hoadley, elasticdesign.co.nz

McFadyen picnic rug | Otto & Spike, ottoandspike.com.au

Handmade concrete planter | Breeana Dunbar Photography, breeanadunbar.com | Mel Evans, melevansphotography.com | Marnie Hawson, marniehawson.com.au

Ebb Tide pebble plates | KW Ceramics, kwceramics.com.au

7" and 12" vinyl records | Connor Dalton, Amanda Harcourt at Zenith Records, zenithrecords.org

Dapple-grey Stallion rocking horse | Olivia O'Connor, oliviaoconnor.com.au

Shell Cordovans | Enrile, enrile.es

Cirus Street skateboard | Cirus Skateboards, ciruskateboards.com

Beard soap | Craftsman Soap Co, craftsmansoap.com

Wood Grain BoomCase sound system | BoomCase, theboomcase.com

Ring-necked parakeet specimen | Jazmine Miles-Long, jazminemileslong.com

Salvage Stool | DesignTree, designtree.co.nz

Traveller Stove | Anevay, anevay.co.uk

The Hydra straight razor | Portland Razor Co., portlandrazorco.com

Grey three-piece window pane check suit | G.A. Zink and Sons, zinkandsons.com.au

TRIIIO tables | Yuta Sawamura, yuta-sawamura.com

Bejewelled Forest Kitsune tattoo | Shaun C. Duncan, shauncduncan.com

Cube Glass whisky tumbler | Nate Cotterman, natecotterman.com | J2 Photography, jsquaredphotography.com | Tyler Barry

Cannon wheel | Greg Rowland, wheelwrightsshop.com

Slipper woodworking plane | Oliver Sparks, oliversparks.co.uk

McGonigle Ceol minute repeater wristwatch | McGonigle Watches, mcgonigle.ch

OTHER ACKNOWLEDGMENTS

Gubbeen Smokehouse Chorizo | How to make chorizo recipe adapted from the book *Gubbeen – The Story of a Working Farm and its Foods* by Giana Ferguson. Published by Kyle Books, 2014

Handmade concrete planter | Breeana Dunbar Photography images used with permission from Etsy Australia

Stoneground wholemeal sourdough bread | Bread by Sprout Bakery

ABOUT THE AUTHOR

Vanessa Murray is a Melbourne-based culture writer and communications nerd. She lives with her furniture maker partner and her son, and is slowly but surely amassing a fine collection of beautiful, practical, long-lasting things. She occasionally endeavours to make some of her own. *Made to Last* is her first book. vanessa-murray.com

THANK YOUS

Big thanks to the team at Hardie Grant – Melissa Kayser, Kate Armstrong, Megan Cuthbert, Vanessa Lanaway, Nick Tapp and Murray Batten – for your interest, direction and sage advice, and for your editing and design smarts. To Anna Carlile, Juliette Elfick, Jane Winning, Matthias Lanz, Rachel Kalmar, Kati Freeman, Jessa Boanas-Dewes and Angus Hervey, thank you for listening to and encouraging me and making some spot-on recommendations. Laura Martinez, Spanish superstar, gracias for your translation time. Matt Martini, thank you for the Japanese translation and maker liaison. Thank you Bridget Hoadley, origami queen. And to Steven Manos, thank you for all of the things but mostly, for taking this leap with me.

Published in 2017 by Hardie Grant Travel, a division of
Hardie Grant Publishing

Hardie Grant Travel (Melbourne)
Building 1, 658 Church Street
Richmond, Victoria 3121

Hardie Grant Travel (Sydney)
Level 7, 45 Jones Street
Ultimo, NSW 2007

hardiegranttravel.com

Explore Australia is an imprint of Hardie Grant Travel

A Cataloguing-in-Publication entry is available from the catalogue of the National Library of Australia at www.nla.gov.au

Made to Last
ISBN 9781741175240

Commissioning editor
Melissa Kayser
Managing editor
Marg Bowman
Project editor
Kate J. Armstrong, Megan Cuthbert
Editor
Vanessa Lanaway
Proofreader
Nick Tapp
Production Manager
Todd Rechner
Typesetter
Megan Ellis
Designer
Murray Batten

Pre-press by Splitting Image Colour Studio
Printed in China by 1010 Printing International Limited